ADVANCES IN CERAMICS • VOLUME 18

COMMERCIAL GLASSES

ADVANCES IN CERAMICS • VOLUME 18

COMMERCIAL GLASSES

Edited by

David C. Boyd
and
John F. MacDowell
Corning Glass Works

The American Ceramic Society, Inc.
Columbus, Ohio

Proceedings of a special symposium on commercial glass held at the 1984 Fall Meeting of the American Ceramic Society, October 17–19, 1984, Grossinger, New York.

Library of Congress Cataloging in Publication Data

Commercial glasses.

    (Advances in ceramics, ISSN 0730-9546 ; v. 18)
    "A special symposium on commercial glass during the 1984 fall meeting of the Glass Division of the American Ceramic Society" —Pref.
    Bibliography: p.
    Includes index.
    1. Glass—Congresses. 2. Glass manufacture—Congresses. I. Boyd, David C., 1924-   . II. MacDowell, John F. III. American Ceramic Society. Glass Division. IV. Series.
TA450.C57   1986   666'.1   86-10723
ISBN 0-916094-78-2

Coden: ADCEDE

ISBN 0-916094-78-2

Coden: ADCEDE

Printed in the United States of America.

Preface

The challenge of glass commercialization encompasses the art and science of formulating an optimum glass composition, discovering and/or creating a need for it in the marketplace, and then manufacturing it profitably. A special symposium on commercial glass during the 1984 Fall Meeting of the Glass Division of the American Ceramic Society brought together technical leaders in all three of these areas of endeavor. This book contains most of the papers from that symposium.

Commercial glass is classified by use into flat, container, fiber, and specialty glass. The flat and container companies produce over ten million tons of soda-lime glass per year. The specialty glass and fiberglass companies melt hundreds of different compositions to produce pressed, blown, and fiberglass for the optical, television, lighting, consumer, communications, and electronics industries. Following a treatise on the history of glass manufacturing, the first section of this book discusses the glass industry as classified by its product.

Glass is a product, but glass is also a material. Because its properties can be tailored to precisely fit its intended use, glass enjoys a unique versatility. The technology of glasses ranges from the ultrapure and low optical loss waveguide materials to the highly strengthenable aluminosilicates to the myriad of phase-separated glasses such as opal glasses, photosensitive glasses, and glass-ceramics. Glass, a most ancient material, has taken its place as one of the leading materials of our modern society. The second section of this book describes the most important of the glass composition types.

Just as our materials scientists invent more sophisticated glass compositions in our laboratories, so must our engineers and process scientists develop more sophisticated methods for the manufacture of these materials in our plants. To do so profitably and competitively is becoming the challenge of the decade. The third section describes some aspects of glass manufacture and the control of the manufacturing process.

We are indebted to the authors who prepared these papers for their willingness to share their expertise with the rest of us in the industry. We hope that this volume will be especially valuable to scientists and engineers in industries peripheral to glass, such as electronics, telecommunications, information, space exploration, and national defense. Although glass-derived materials are used only sparingly in products from these industries, the unique properties of glass often constitute a most vital element in the operation of their complex product systems. We also wish to thank the American Ceramic Society for affording us the opportunity to publish this collection of knowledge on commercial glasses.

David C. Boyd
John F. MacDowell
Corning Glass Works

Contents

SECTION III: GLASS MANUFACTURE

3500 Years of Glassmaking

GEORGE B. HARES

Corning Glass Works
Corning, NY   14831

The evolution of commercial glassmaking from its beginnings in ancient Egypt to present-day practice is traced. The relationships between raw material availability, changes in melting and forming processes, expanding technological needs, and glass chemistry are emphasized.

As scientists involved in commercial glass manufacture, we are possessors of a long and rich technological heritage, reaching back some 3500 years. This paper is only a brief and cursory overview of the evolution of commercial glassmaking from its beginnings to present-day practice.

The origins of glassmaking are lost in antiquity. Dates as early as 2600 B.C. have been claimed for some glass beads, but at least some of these dates are questionable.

The immediate predecessor of glass may have been a material known as Egyptian faience, which was used to make small objects such as statuettes or jewelry. This material appeared as early as 4000 B.C. and was prevalent throughout the Egyptian empire from 2700 to 350 B.C. It consisted of a porous, silica body with a surface of alkali silicate glass, often colored green or blue. This surface probably was not produced by a classical glazing technique, but was rather the result of chemical reactions with a mixture of sand and alkali salts in which the object was buried during firing. The maximum temperature during firing was probably no higher than 900°C, judging from X-ray evidence of the crystalline phases of silica present in the body.

Faience had been manufactured for at least a thousand years before the earliest glass objects appeared, in the form of "sandcore" glass vessels used as containers for cosmetics, and so forth. Although these glasses were most often deep blue (probably an imitation of lapis lazuli), the base glass composition differed little from lime-glass compositions used in the flat-glass industry today. While analyses of the ancient compositions show them to be very complex mixtures, actually only a few readily available raw materials were probably used: Sands in the Nile valley contain a large amount of limestone, and either natron or plant ash, both of which contain considerable quantities of lime and magnesia, as well as alkali, was used for fluxes.

The melting process was probably carried out in two stages: First, at comparatively low temperature, the sand and alkali were converted into a frit; and second, the frit was converted to glass in crucibles in the melting furnace. The highest fritting temperature was probably about 750°C, while the melting temperature was no higher than 1100°C and the practical temperature was about 1050°C. (If pieces of the crucibles are heated to 1150°–1200°C, they themselves form a glass!) At 1050°C these compositions were very viscous, so that forming a glass vessel was a problem. Unfortunately, no written record of Egyptian glass-forming technique survives; thus, the technique has been subject to much speculation over the years.

Since most of the vessels show traces of a sand or clay core material adhering to the inside surface, we must assume that the vessels were formed around a core. This core was possibly preshaped and, by some process, dipped or rolled in the softened glass. It was decorated by applying threads of colored opaque glass, usually yellow or white, which were then rolled flat into the surface. Reheating the decorated vessel and then "combing" the surface with a pointed tool produced more elaborate decorations. Opaque glasses were probably obtained by adding a white pigment (calcium antimonate) or a yellow one (lead antimonate) to the base glass used for the body of the vessels.

In any event, glass objects in this historical period were available only to the wealthy or powerful. It was not until the Roman period and discovery of the blow pipe, to make thin-walled vessels, that glass products became more universal. The introduction of the blow pipe may have resulted from new techniques for achieving higher glass-melting temperatures: As iron implements became more widespread, higher smelting temperatures than those used for copper smelting became mandatory. These newer methods of glass melting allowed temperatures at which the glass was fluid enough to gather on a hollow rod, so that a glass bubble could be formed by blowing through the tube.

The blow-pipe technique made possible the rapid production of utilitarian vessels, and glass became a household commodity. A middle-class Roman family probably owned glass storage containers, looked through crude glass windows, and could even buy glass souvenir cups of their favorite gladiator. Everywhere the Romans went, glass went, too. Glass objects from Roman times have been found in abundance to the ends of the empire, as far north and west as Scandinavia and Britain, and as far east and south as eastern Syria and Ethiopia.

These early glasses used two primary raw materials, sand and plant ash, neither of which had a fixed chemical composition. Sand varied in content of iron, alumina, lime, and other impurities. Plant ash, derived from burning of various plants or trees, varied widely in constitution, reflecting not only the type of plant, but also the soil in which the plant grew: For instance, plants that grew near the sea or on salt desert land had a high proportion of soda to potash, while inland plants were much richer in potash than in soda. With universal availability of the chief glassmaking raw materials, sand and ash, it is easy to see that growth and decline in glass technology closely paralleled growth and decline of the individual civilizations and empires. In addition, the easy workability of glass as an artistic medium resulted in forms very characteristic of the locale or of the origin of the craftsman himself.

After the fall of the Roman Empire, the main glassmaking centers returned to the eastern Mediterranean area. Islam was on the rise, and enameling and copper- and silver-staining techniques were applied to produce beautifully decorated mosque lamps.

During the Middle Ages (A.D. 500–1500), glassmaking in central Europe declined, but did not disappear completely. The glassmakers were itinerant, setting up their furnaces near a supply of sand and a forest. The wood from the forest supplied two needs: fuel for the furnaces and ashes, which could be leached to provide raw material for the glass batch. Because the impurities in the sand, particularly iron, were very high (compared to the sand of the Mediterranean beaches), this glass had a characteristic dark green, almost black, color. Glass of this type has come to be known as "Waldglas" or forest glass. Since the forest was quickly depleted, these glassmakers then gathered up all their necessary tools and moved on to another spot, where the supply of sand and wood was adequate.

Following the Crusades, and with the beginning of the Renaissance, the artistic glassmaking pendulum swung back from Asia Minor to the western Mediterranean, particularly to Venice. Not only was this a period of great artistic activity, but it was also a period in which technical emphasis shifted to obtaining ever purer raw materials, resulting in the almost colorless, transparent Venetian "cristallo" of the sixteenth century.

The traditions of luxury glassmaking, carried by Italian craftsmen, spread through Europe like wildfire in the sixteenth and seventeenth centuries. The Venetian tradition, established in England by 1575, remained more or less active through the Puritan period and into the Restoration (1660). At that time, in England, there was a great effort toward establishing national self-sufficiency. Since most glassmaking ingredients were still imported, in 1673 George Ravenscroft was engaged by the Glass-Sellers Company to begin research on eliminating foreign ingredients. By about 1676 he had developed a potash lead glass that was insusceptible to atmospheric moisture. Ravenscroft was permitted to use a raven's head seal as his device on glassware of this composition. For silica, English flints, a very hard form of quartz, low in iron, were used. The flints were pulverized before mixing with the potash and lead, and these potash lead glasses became known as flint glasses due to their color clarity.

There were two basic processes for making window glass: the cylinder process and the crown process. In the cylinder process, a long, cylindrical bubble was blown and the end heated, then opened to form a cylinder. The entire piece was next reheated, slit lengthwise, and opened out and allowed to sag into a flat sheet. This process was used (with some mechanization) until the beginning of the twentieth century.

The crown process, on the other hand, consisted of first forming a spherical bubble. A pontil rod was then attached to the surface of the sphere, diametrically opposite the blowpipe. The blowpipe was cracked off and the sphere reheated and opened where the blowpipe had been attached. Then the sphere was reheated, and when the glass fluidized the pontil rod and sphere were rotated very rapidly. Centrifugal force opened the sphere into a more-or-less flat disk. The round sheet or "table" was then cracked off, annealed, and cut up for window panes. In ninteenth-century England, sheets made by this process were as large as five feet in diameter.

Two characteristics of glass made by the crown process were a thickness that decreased with increasing distance from the center, and a large mass of clear glass, known as a bull's-eye, or crown, in the center of the "table," where it had been attached to the pontil iron. The bull's-eyes themselves were shaped somewhat like lenses; indeed, in sixteenth-century Venice, spectacles were made using bull's-eye glass.

The seventeenth century brought great activity in the science of astronomy. Galileo built his first telescope in 1609, and larger and larger telescopes were built for higher magnifications. It was difficult, however, to obtain sharp images, since eliminating the color fringes at the edge of the image seemed impossible. Then in 1666, Newton discovered that a beam of sunlight passing through a glass prism would bend and form the different colors of the spectrum. He concluded that "all refractive substances diverged the prismatic colors in a constant proportion to their mean refraction" and drew the natural conclusion that refraction could not be produced without color.

Newton's assumption, however, was incorrect. Chester Moor Hall argued that the humors of the eye refract rays of light to produce a color-free image on the retina, and he argued reasonably that combining lenses of different substances might produce a similar result. After some time, he discovered a combination of two glasses that would indeed accomplish this purpose. This achromatic lens was perfected by John Dollard in 1758, using a combination of window (crown) glass for the positive component and lead-crystal (flint) glass for the negative component. The terms "crown" and "flint" are used in optical-glass designations to this day. This combination of a convex lens of crown glass and a concave lens of flint glass decreased the color fringes on the image formed by a telescope, but the colors did not disappear completely: A much smaller, colored edge, the so-called secondary spectrum, remained.

In the early nineteenth century, chemists began introducing oxides other than lead and lime into glass, hoping to discover new types of optical glass. Fraunhofer had determined that, if chromatic aberrations were to be eliminated, new glasses, with different ratios of partial dispersion than the ordinary crown and flint glasses, would have to be found. Most of the search, however, was fruitless, because of difficulty in obtaining experimental melts suitable for optical measurements.

Not until Otto Schott (who was perhaps the first chemist to present a doctoral dissertation on a glass-related subject) approached Ernst Abbe in 1879, asking him to determine the optical properties of an experimental lithium glass he had made, did the quality of small glass melts become acceptable. Schott joined forces with Abbe, and by 1884 the Jenaer Glasswerke Schott and Genossen was formed. By 1886, 44 glass types were on the market. Among the new glass types were phosphate and borate crowns, barium silicate crowns, borosilicate crowns, zinc and potash silicate crowns, and borate flints. Minor variations proliferated over the years. One significant advance was the development of the high-index, low-dispersion glasses based on lanthanum borate.

Along with advances in optical-glass compositions came continued improvement in the quality of optical-glass melts, but this aspect of optical glass must be discussed elsewhere. Other Schott developments included glass compositions for thermometry and borosilicate compositions for chemical ware.

The first commercial laboratory in the United States devoted to glass research was founded at Corning Glass Works in 1908. One of its first problems was the thermal shock breakage of railroad lantern globes and signal lenses. A soft, lead-containing, low-expansion borosilicate glass was developed for this purpose. A clear version, known as Nonex,* was used for battery jars because of its chemical stability. The bottom of one of these battery jars was used to bake the famous cake that resulted in the use of borosilicate glass for consumer baking ware.

At this same time, World War I cut off the supply of German borosilicate chemical ware, and a new borosilicate glass was developed to meet the need. Lower in expansion and much more difficult to melt than Nonex, this composition has now become the standard for laboratory ware around the world. One of the characteristics of the borosilicate family is the wide range of properties available using different proportions of alkali and boric oxides: Extremely low-expansion glasses such as those used in the 200-inch Mt. Palomar telescope, moderate-expansion glasses for sealing to molybdenum and Kovar,[†] and higher-expansion glasses such as those used in optical borosilicate crowns are typical illustrations of this family.

---

*Corning Glass Works, Corning, NY.
[†]Westinghouse Electric Co., Pittsburgh, PA.

4

Among the problems associated with borosilicate glasses is the tendency of many such glasses to separate into a silica and a borate phase, the latter having poor chemical durability. This tendency, however, has been capitalized on by adjusting the composition so that the separated borate phase is an interconnected network. Subsequent leaching of this phase in dilute acids, followed by firing at about 1250°C, produces a hard, clear glass, more than 96% silica, that is marketed and known in the trade as Vycor.[‡] It has a low thermal expansion (8×10^{-7}/°C) and can withstand much higher temperatures than can the borosilicate family.

In the nineteenth century, two families of glass were widespread: The lime-glass family was used for windows, bottles, dinnerware, and such decorative ware as vases. Some lime glasses were called "flint" glasses, for their color clarity (low iron content), rather than their optical properties. The other family was the alkali-lead silicate family. The alkali was most often potash, and two lead-oxide contents were prevalent, one at $\approx$24 wt% lead oxide, the other at $\approx$30 wt%. These compositions were used in the cut-glass industry.

Edison's first light bulbs were made from lead-glass tubing, and the practice of using lead glass for lamp envelopes continued until World War I, when the supply of lead oxide was severely curtailed. A switch was made to a lime-glass composition containing about 8 wt% dolomitic lime, in contrast to the 10–12 wt% lime used in the container and window-glass industries. Later, tubing for fluorescent lamps was made from the same composition.

The search for glasses compatible with high-temperature applications, where the borosilicate family would be unsuitable, led to development of the aluminosilicate family. This family was originally used for combustion tubes. The compositions were essentially alkaline earth aluminosilicates, with a small amount of alkali. While the annealing point was about 150°C higher than that of the low-expansion borosilicates, the viscosity–temperature relationship was much steeper, so that the glasses could be melted in conventional units. High annealing and strain points permitted tempered items of these glasses to be used for top-of-the-stove cooking. Other aluminosilicate glasses include the high-soda aluminosilicates, which are chemically strengthened by a potassium for sodium-ion exchange below the strain point, and the alkaline earth glasses used in the fiberglass industry.

The ultimate in low-expansion and high-temperature capability is vitreous silica itself, which can be prepared by heating silica and/or quartz crystals to above the melting point of silica (>1725°C). However, the glass is so viscous that good quality vitreous silica is extremely difficult to obtain in this manner. An alternative technique is a vapor-deposition process, in which silicon tetrachloride is reacted with oxygen above 1500°C. A finely divided, particulate vitreous silica is formed, which can be consolidated on a substrate above 1800°C. This procedure (developed in the 1930s) is the basis for the processes used to make vitreous silica optical waveguide fibers, because of the extreme purity achievable by the vapor-deposition process.

The evolution and development of glass compositions and forming methods for television picture tubes would require a separate paper, but the original composition used in small black and white tubes was a potash lead silicate, followed rather quickly by an alkali barium glass with very similar physical and electrical properties. The advent of color television required development of a higher-

---

[‡]Corning Glass Works.

expansion barium glass for the panel, a lead-containing glass for the funnel, the original alkali lead glass for sealing to the electrical leads, an intermediate-expansion lead glass for the neck, and a solder-glass frit to seal the panel and funnel together after the phosphors, etc. were processed. Subsequent high-voltage requirements resulted in substitution of strontium oxide for barium oxide to increase the X-ray absorption of the substrate.

Opal glasses have been common since the earliest times. Early opacifying agents were calcium antimonate (white), lead stannate (yellow), and cuprous oxide (red). Phosphates and arsenates came into use much later. Fluoride opals first came into use in about the middle of the nineteenth century.

What can be said about colored glasses? They have been with us since the beginning of glass technology. Indeed, it seems that every time a chemist (or alchemist) discovered a new material, he tried it in a glass-composition matrix to see what color he would get. The story and evaluation of colored glasses is beyond the scope of this paper.

A variation of colored glass is the silver- and/or gold-containing glasses which respond to ultraviolet light. Among these are the photosensitive clear glasses; the photosensitive opal glasses, where silver particles act as nuclei for a crystallization phase; glass-ceramics, either photonucleated or internally nucleated; and the photochromic silver-halide–containing glasses.

As glass scientists, we must continually realize that glass is an extremely versatile material with a long history, an exciting present, and a promising future.

Section **I**

The Glass Industry

Optical Glasses

EMIL W. DEEG

AMP Incorporated
Harrisburg, PA   17105

Material properties, critical for meeting performance requirements of passive and active optical elements in specific applications, and important manufacturing aspects are discussed. A brief historic review, primarily of glass composition development, is included.

Optical glasses are the first of what we call today "high-technology materials." Since their introduction as regular commercial products more than a century ago, they have had to meet exceptional and continuously increasing requirements for actual values and melt-to-melt reproducibility of physical properties, as well as quality and forms of supply. High transmittance in the visible spectrum, or, in the case of optical filter glasses, well-defined "color," freedom from cords, striae, bubbles, seeds, and solid inclusions are synonymous with optical-quality glass. Producing such materials requires a thorough knowledge of the quantitative relationships between composition and properties. History shows that optical-glass production also required first the skills and later the technology for monitoring and controlling both reliably and precisely the melting, forming, and annealing processes.

This paper attempts to introduce optical glasses, emphasizing properties important for optical instruments. Compositions, given in round numbers only to demonstrate their variety, represent mature glasses available in commerce for more than 20 years. Not considered are fiber optics, glass-ceramics for reflective optics, secondary manufacturing operations (i.e., grinding, curve generating, polishing, slumping/sagging, toughening, surface modification by chemical treatment, and vacuum deposition of thin films), use of optical glasses for nonoptical purposes (e.g., ultrasonic delay lines), optical cells for laboratory apparatus, etc., or illuminating and lighting optics. Results of recent glass compositional research, although interesting, are not included, for reasons implicit in the historical highlights.

An "optical glass" must meet certain minimum requirements, i.e.,

1. Identifiability by the customary set of optical properties.
2. Satisfactory chemical durability.
3. Manufacturability in acceptable quality and in pieces large enough for preparing useful optical elements.

Quality of the highest level compatible with performance and cost is a concern also for ophthalmic glasses, which usually are not considered "optical." In that case, the consumer benefits directly from the vast knowledge and expertise accumulated by the optical-glass industry. In addition, the manufacture of filter glasses and the preparation, in reasonable sizes, of the more advanced amorphous materials, such as laser glasses, used in optics would be impossible without optical-glass technology.

Historical Highlights

This review highlights the major technical events leading to the optical-glass industry as we know it today. For details, especially regarding commercial aspects, see Refs. 1–9. More recent acquisitions of products and product lines also have been reported in trade and business journals.

At the beginning of the eighteenth century, dispersion was considered a property of light, and not of an optical material.[10] Although flint glass was about to emerge, Newton and his scientific contemporaries recognized only one "glass," the alkali-lime-silicate crown. Its composition was determined by locally available raw materials and the traditional, jealously guarded glassmaking recipes. Melting technology was still essentially as described in Agricola's *De re metallica* (ca. 1500).

Publications of Neri in 1612 and 1681 and by Kunckel in 1679 and 1689 opened the field of glass chemistry considerably but had no immediate, direct effect on optical-glass development. Early in the nineteenth century, Joseph von Fraunhofer needed improved glasses for his optical instruments and experiments; at that time it was already known that a combination of crown and flint lenses reduced the chromatic aberration of a lens system. Fraunhofer, working with Utzschneider, the owner of a glass factory, not only found that the chemical durability of a glass depends on its composition, but also prepared a new crown glass. This glass, in combination with flint glass, permitted a better correction for chromatic aberration than did the regular crown available at the beginning of the nineteenth century. Pierre-Louis Guinand, who introduced mechanical stirring of the glass melt, also worked with Fraunhofer and Utzschneider.

In 1824 the Royal Society elected a commission charged with experimental studies of flint glasses. The members were Michael Faraday, John Dollond (optician), and Sir John F. W. Herschel (astronomer). In Faraday's laboratory a lead-borosilicate flint was prepared with refractive indexes of 1.8521 in the red spectrum, 1.8735 in the yellow, and 1.9135 in the violet. With available equipment, no large high-quality pieces could be made, and melting still used batch fritting, although fining was done in platinum.

The most radical step away from conventional nineteenth-century glass chemistry was taken by William Vernon Harcourt[11] in the first half of that century. For studying the effect of different glass constituents on optical properties, he designed a laboratory "furnace" permitting glass melting in air, free from the effects of combustion gases above the melt surface. This device consisted of a rotating platinum crucible heated directly from the outside by a hydrogen-fed ring burner with platinum nozzles. The conical crucible was not contained in a furnace. In cooperation with G. G. Stokes, Harcourt used this device to explore the effect of glass composition on optical properties. The impressive list of new glass constituents introduced by Harcourt during these experiments was published in 1871 by Stokes.[12]

Biological and medical scientists needed a wider selection of new, high-quality glasses for optical instruments to build microscopes with improved resolution. By the last quarter of the nineteenth century microscopic theory had developed enough to permit quantitative specification of what were, at that point, hypothetical glasses. The status of the optical-glass industry, as seen in 1876 by its prime "customer," the instrument designer, was summarized by Ernst Abbe:

> ...some experiments in the production of glasses with small secondary dispersion conducted by Stokes in England a few years ago, though barren of direct practical result, gave useful hints as to the specific effects of certain

bases and acids on refraction of light. The uniformity shown by existing glasses in their optical qualities is probably chiefly due to the very limited number of materials hitherto used in their manufacture. Beyond silicic acid, alkali, lime and lead, scarcely any substances have been tried, except alumina and thallium. When this narrow groove is left, and a methodical study, on an extended scale, is made of the optical qualities of chemical elements in combination, we may anticipate with some confidence a greater variety in the products.[13]

In the same publication, Abbe also very clearly assigned the responsibility for progress in optical insturmentation:

The future of the microscope as regards further improvement in its dioptric qualities seems to lie chiefly in the hands of the glassmaker.

Notice that the call for action addressed not the chemist or scientist but the "glassmaker," i.e., the glass technologist.

In one of the rare events in the history of technology, the right person, endowed with the necessary talents, knowledge, and experience, developed at the right time an active interest in the problem that Abbe had outlined: Otto Schott, son of a glassmaker, was educated as a chemist, interested in high-temperature chemistry, and had access to a glass-melting laboratory; he was also gifted with vision, the strength to insist on excellence, and an ability for leadership and teamwork. Schott contacted Abbe in 1879 and a joint investigation began. Schott's previous experience in organizing both a chemical and a glass plant in Spain became an additional, valuable asset. The results of their efforts did not remain in the laboratory, but were carried through small-scale manufacturing to the market place.

The first catalog of the new enterprise was published in July of 1886. Its introduction acknowledged that the costly experiments on a manufacturing scale had been subsidized liberally by the Prussian government and by Carl and Roderich Zeiss, manufacturers of optical instruments. The combination of government seed money and total financial commitment by Schott, Abbe, and the Zeisses led to commercial realization of the scientific work and resulted in a timely introduction of the desired selection of high-quality glasses for the emerging optical industry. It is interesting to note that the 1886 catalog, with supplements in 1888 and 1892, listed 76 glasses, 6 of them compared to optical glasses already on the market ("exactly corresponds to Chance's Hard Crown" or "optically identical with Chance's Extra Dense Flint," etc.). In addition, 7 of the 76 were marked specifically "to be used under protection" or by a similar warning indicating low chemical durability.

Table I highlights the early history of the approximately 250 optical glasses available today. The introduction of lanthanum oxide as a glass ingredient by G. W. Morey[14] extended the range of useful optical glasses. Fluoride glasses, suggested by K. H. Sun,[15] offered another extension, but they could not be used because of low chemical durability.

Process development since the turn of the century was carried out by teams of scientists and engineers working for companies in France, Germany, the United Kingdom, the United States (after World War I), Japan (since about 1935), and the U.S.S.R. (after World War II). It followed the advancement of melting and forming techniques in the glass industry in general. To meet increasing demand for refractive index control, special emphasis was placed on precision annealing. Continuous melting was generally accepted in the optical-glass industry only after World War II. Its adaptation required size reduction of the tanks, increase in the

11

Table I. Highlights of History of Glass Chemistry and Melting Technology Important for the Development of Optical Glasses.

Year	Elements Introduced as Glass Ingredients	Technological Development Crucial for Progress in Optical-Glass Preparation
1689	Kunckel: Na, K, Ca, Sn, Pb, B, Si, P, As; O, S; (Cu, Au, Mn, Fe, Co)	
1803		Guinand: mechanical stirring of melt
1817	Fraunhofer: Bi	
1829	Doebereiner: Sr, Ba	
1830		Faraday: fining in platinum
1844		Harcourt: melting in platinum; combustion gas-free atmosphere above melt
1851	Maes: Zn	
1866	Lamy: Tl	
1871	Harcourt: Li, Be, Mg, Cd, Al, Ti, Sb, V, Mo, W, U; F; (Ni, Cr)	
1900	Schott (Jena Glass Lab): Hg, Nb, Ce, Didymium, Er, (Ag).	Special refractories and melting conditions for various glasses; fused silica lens

Only elements not introduced previously are listed for the particular period. Fraunhofer's invention of the spectrometer made possible precision measurements of refractive indexes of glasses.

length of fining and conditioning zones relative to the melter, development of sophisticated stirring techniques and temperature controls. Most importantly, the bold decision was made to line forehearth and feeder with platinum. Electric booster heat and direct melting are now common in the mass production of optical and ophthalmic glasses.

Optical Properties

Refraction and Dispersion

Refractive index measurements on optical glasses are based on Snell's law

$$n = \sin i / \sin r \tag{1}$$

with n = refractive index of glass vs vacuum, $n = n(\lambda)$; i = angle of incidence; r = angle of refraction; and λ = wavelength of light.

Optical glasses today have refractive indexes of ≈ 1.44–1.99 in the blue spectrum and 1.43–1.94 in the red. Knowledge of the refractive index as a function

12

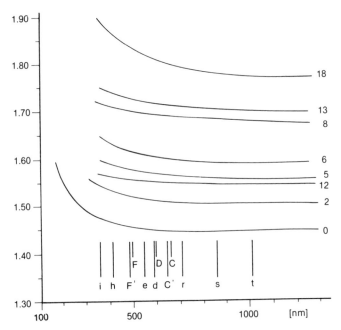

Fig. 1. Dispersion of various optical glasses. Curves are identified by glass numbers of corresponding nominal compositions given in Tables V and VI; 0 = fused silica. Locations of selected Fraunhofer lines are indicated in the lower portion of diagram (Table II).

of wavelength (Fig. 1), that is, the "dispersion" of a glass, is essential for the lens designer concerned with reducing chromatic aberration: Glass data sheets therefore list refractive index values for up to 18 wavelengths. In addition, interpolation formulas for approximate wavelengths of 365–1000 nm are given in the catalogs. The dispersion formula most frequently used today is a power series $n = \Sigma_{i=0}^{5} C_i \lambda^{(2i-8)}$. Constants C_i, for interpolation to approximately $\pm 5 \times 10^{-6}$, are included in the data sheets of most optical glasses.

Glass Dispersion and Chromatic Aberration of a Lens System

Just as the index of refraction is a function of the wavelength of light, the optical properties of a refractive system depend also on its wavelength. The thin lens may serve as an example of the consequences of this relationship,

$$\phi = 1/f = (n_g - n_a)(1/R_1 - 1/R_2) \tag{2}$$

Here ϕ = lens power; f = focal length; n_g = refractive index of glass; n_a = refractive index of the environment, usually air; and R_1 and R_2 = radii of curvature of lens surfaces.

Logarithmic differentiation of Eq. (2) yields, with the substitutions $n_g = n$, $n_a = 1$:

$$df/f = -d\phi/\phi = -dn/(n-1) \tag{3}$$

The differential dn or, for practical purposes, a suitable difference $\Delta n_{x,y} = n_x - n_y$ of refractive indexes, measured at the wavelengths x and y, characterizes the

Table II. Correlation Between Fraunhofer's Alpha Notation and Wavelength of Frequently Used Spectral Lines

alpha Notation	Wavelength (nm)	Spectral Range	Light source: Gas Discharge in
i	365.0146	Ultraviolet	Mercury vapor
h	404.6561	Violet	Mercury vapor
g	435.8343	Blue	Mercury vapor
F'	479.9914	Blue	Cadmium vapor
F	486.1327	Blue	Hydrogen
e	546.0740	Green	Mercury vapor
d	587.5618	Yellow	Helium
D	589.2938	Yellow	Sodium vapor
C'	643.8469	Red	Cadmium vapor
C	656.2725	Red	Hydrogen
r	706.5188	Red	Helium
s	852.1101	Infrared	Cesium vapor
t	1013.98	Infrared	Mercury vapor

The sodium D line represents the center of the doublet. Other light sources used for index measurements include helium-neon lasers (632.8 nm) and neodymium glass lasers (1060.0 nm). Mercury gas-discharge lines are used for measurements in the far infrared (1529.6, 1970.1, and 2325.4 nm).

dispersion of a glass. It is general practice in the optical industry to use Fraunhofer's designation of spectral lines instead of wavelength subscripts (Table II).

A correlation between the wavelength dependence of the lens power and a material constant can be derived from Eq. (3),

$$ d\phi/\phi = 1/v_z = (n_x - n_y)/(n_z - 1) \qquad (4) $$

where v is the Abbe value (also Abbe number, constringency, constringence value, reciprocal dispersion, reciprocal dispersive power, v value) of the glass.

The wavelengths x, y, and z of Eq. (4) are chosen to yield

$$ v_d = (n_d - 1)/(n_F - n_C) \qquad (5) $$

$$ v_e = (n_e - 1)/(n_{F'} - n_{C'}) \qquad (6) $$

The original Abbe value $v_D = (n_D - 1)/(n_F - n_C)$ is still used occasionally. Today's optical glasses cover $\approx 20 < v_d < 90$.

Within the spectral range F (blue) to C (red), if $d\phi/\phi \neq 0$, the focal length of a lens differs for the various wavelengths, and colored edges of an object's image are produced with white light ("chromatic aberration"). Failure to correct this aberration would be most obvious to the general public by its effect on photographic performance: Sharp pictures would be impossible under the most common conditions.

Equation (2) determines the power of a lens at a given wavelength and Eq. (3) the relative change of the lens power $d\phi/\phi$ with wavelength or "color" of light. Together, the two material constants n_z and v_z describe the contribution of an optical glass to refractive power and chromatic aberration in a lens of given geometry. It therefore became customary to characterize an optical glass by ordered pairs $\{n_D; v_D\}$ and later, when light sources with better-defined wavelengths became available, $\{n_d; v_d,\}$ $\{n_e; v_e\}$. Plotting these ordered pairs in an $\{n; v\}$ plane results in a map of optical glasses (Fig. 2).

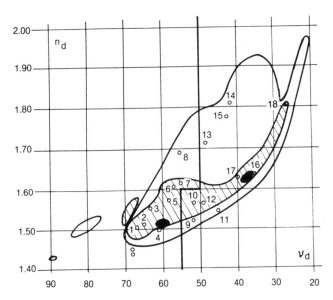

Fig. 2. Representation of optical glasses in the $\{n_d; v_d\}$ plane. Glasses listed by numbers correspond approximately to nominal compositions given in Tables V and VI; 0 = fused silica. Solid dividing line separates crowns (left) and flints (right). Diagram also contains historically significant boundaries: solid areas = 1870, hatched area = 1920, solid boundaries only = 1984.

The index difference $n_F - n_C$ expresses the *total dispersion* of a glass from the blue F line to the red C line of the spectrum. Textbooks on lens design and classical optics (e.g., Refs. 16–20) show that index differences for narrower intervals of the spectrum are required for computing and correcting more accurately the achromatism of a lens system. These *partial dispersion* values are expressed as fractions of the total dispersion and called *relative partial dispersion*, $P_{x,y}$. Examples are

$$P_{s,t} = (n_s - n_t)/(n_F - n_C), \qquad P_{C,s} = (n_C - n_s)/(n_F - n_C)$$

$$P_{d,C} = (n_d - n_C)/(n_F - n_C), \qquad P_{e,d} = (n_e - n_d)/(n_F - n_C)$$

$$P_{g,F} = (n_g - n_F)/(n_F - n_C), \qquad P_{i,h} = (n_i - n_h)/(n_F - n_C) \qquad (7)$$

$$P'_{s,t} = (n_s - n_t)/(n_{F'} - n_{C'}), \qquad P'_{C',s} = (n_{C'} - n_s)/(n_{F'} - n_{C'})$$

$$P'_{d,C'} = (n_d - n_{C'})/(n_{F'} - n_{C'}), \qquad P'_{e,d} = (n_e - n_d)/(n_{F'} - n_{C'})$$

$$P'_{g,F'} = (n_g - n_{F'})/(n_{F'} - n_{C'}), \qquad P'_{i,h} = (n_i - n_h)/(n_{F'} - n_{C'})$$

Such P values range from ≈ 0.53 to 0.63 in the blue spectrum and 0.44 to 0.57 in the red. The $\{P_{x,y}; v_d\}$ diagrams are included in optical-glass catalogs.

For most glasses a linear correlation exists between the relative partial dispersion and the Abbe values v_d (or v_e).

$$P_{x,y} = p_{x,y} + q_{x,y}v_d \qquad (8)$$

where the empirical constants $p_{x,y}$ and $q_{x,y}$ depend only on the wavelengths x and

15

y. Discussion of the power formula of, for instance, a doublet of lenses of different glasses shows that if both glasses obey Eq. (8) achromatization can be achieved for two wavelengths only. To reduce the remaining chromatic aberration ("secondary spectrum"), correction for more than two wavelengths is necessary. This correction requires at least one glass in the doublet to deviate from the "normal," linear behavior according to Eq. (8). Glasses with such anormal relative partial dispersion are identified in the catalogs, and the deviations from the linear relationship

$$\Delta P_{x,y} = (P_{x,y})_{actual} - (p_{x,y} + q_{x,y}v_d) \tag{9}$$

are listed in the data sheets. For available optical glasses, ΔP values range from about $(-1 \text{ to } +4) \times 10^{-2}$ at the blue end of the spectrum and $(-6 \text{ to } +2) \times 10^{-2}$ at the red end.

Notation

The values for n_d and v_d also provide a numeric code of glass identification. For this purpose, the refractive index is carried to three decimals only, the Abbe number to one, and the following six-digit expression is formed:

$$(n_d - 1) (v_d \times 10)$$

Thus, a glass with the catalog values $\{n_d = 1.51680; v_d = 64.17\}$ carries the identification 517 642. Historically, optical glasses are identified by alphanumeric codes such as FK 5, PK 51, K 3, KzFS N4, BSC 510644, TF 530512, DEDF 728284, BaCED 20, ADC 1, BCD C13–58, or FDD C48–34. These designations can mean more than the numeric code: If, for example, the term BK 7 is mentioned, an experienced optical scientist or lens designer recalls that this glass has $n_d = 1.517$ and $v_d = 64.2$, density ≈ 2.5 g/cm^3, is readily available at a reasonable price, has been made in high-quality, $\approx$1-ton pieces, is of excellent chemical durability, and so forth. The term 517 642 describing the same glass covers only the first two statements. An interpretation of some codes of the most frequently used system is given in Table III.

External Effects on Refractive Indexes

Among parameters other than λ affecting the refractive index of a glass, the following are important for optical applications: temperature, stress and strain, external magnetic fields, and intensity of the light beam passing through the glass.

Index changes with temperature $\Delta n/\Delta T$ are $\approx 10^{-6}\,°\mathrm{C}^{-1}$ or $10^{-6}\,\mathrm{K}^{-1}$. It is thus necessary to consider the temperature changes in the refractive index of air against which the index of the glass is measured and used. Therefore, two sets of data describing the temperature dependence are given: (1) the relative index change $(\Delta n/\Delta T)_{rel}$, measured vs air and (2) the absolute index change $(\Delta n/\Delta T)_{abs}$, representing the glass index vs vacuum.

The latter temperature coefficient is not measured directly but computed from the temperature dependence and the known refractive index of air at given pressure, temperature, and humidity.

Low-index glasses usually also have low temperature coefficients: For instance, for a 487 845 glass,

$$(\Delta n/\Delta T)_{rel} = -6 \times 10^{-6}\,°\mathrm{C}^{-1} \tag{10}$$

$$(\Delta n/\Delta T)_{abs} = -7 \times 10^{-6}\,°\mathrm{C}^{-1} \tag{11}$$

For high-index glasses, the temperature coefficients are also high, as illustrated by a 918 215 glass:

$$(\Delta n/\Delta T)_{rel} = 23 \times 10^{-6}\,°\mathrm{C}^{-1} \tag{12}$$

Table III. Some Abbreviations Used for Identification of Optical Glass Groups

Full Name of Glass Groups ("Families")	Abbreviation
Fluor crown	FK
Dense phosphate crown	PSK
Borosilicate crown	BK
Crown	K
Barium crown	BaK
Dense lanthanum crown	LaSK
Extra-light flint	LLF
Flint	F
Dense barium flint	BaSF
Short flint	KzF
Lanthanum flint	LaF

Although some of the expressions contain chemical terms, they are not any more indicative for actual glass compositions. For instance, a glass of the nominal composition $12B_2O_3$-$55SiO_2$-$5Na_2O$-$4K_2O$-$22BaO$-$2Al_2O_3$ could carry the group designation PSK. The maximum number of group designation at present seems to be 27. Another company uses only 22 groups to classify approximately the same number of glasses.

$$(\Delta n/\Delta T)_{abs} = 21 \times 10^{-6}\,°C^{-1} \qquad (13)$$

Both examples are for the blue end of the spectrum, important since the $\Delta n/\Delta T$ values are wavelength-dependent, generally decreasing with increasing wavelength. Within $\approx 20°C$, refractive index as a function of temperature can be described by a linear function

$$n(T) = n_0\{1 + (1/n_0)(\Delta n/\Delta T)T\} = n_0(1 + bT) \qquad (14)$$

With lens systems designed for widely varying temperature conditions, e.g., from -50 to $+80°C$, one must consider the temperature dependence of the $\Delta n/\Delta T$ values; a nonlinear function $n(T)$ must be introduced. For some glasses, the required data are listed in the data sheets. If the data are not given, the glass supplier is usually equipped to measure them on request.

Strain or stress applied to a glass results in birefringence, which can be caused by improper annealing (permanent strain) or by temporary external influences such as temperature variations or improper mounting of the optical element. For massive optics, the volume forces acting on the "lower" parts of the glass piece can also introduce strain. Optically measured birefringence (optical retardation) and mechanical stress are connected by Brewster's law. The stress-optical constants of optical glasses can range from about -1.5 to $+3.6$ brewsters; values for some high-index flints (Pockels' glass) approach zero.

Index changes measured in a strained piece of glass with light polarized parallel to the principal stresses are high for high-lead glasses (high refractive index). The birefringence, however, which is the difference of the index values measured in these two directions, can be very low. On the other hand, low-index glasses show a low absolute index change with applied stress but a high birefringence.[21]

Occasionally the Faraday effect observed on optical glasses is used, for instance in magneto-optical shutters. Among optical glasses, the high-lead-containing dense flints have the highest Verdet constants. At room temperature, in the blue spectrum, they can reach ≈ 0.2 min angle/(G·cm), but decrease rapidly with increasing wavelength to ≈ 0.08 min angle/(G·cm) at the helium-neon laser wavelength (632.8 nm). At this wavelength, certain terbium oxide-containing Faraday rotator glasses assume values of 0.25 and, at the blue end of the spectrum, 0.5 min angle/(G·cm). An additional advantage of the terbium oxide-containing glasses is the relatively low wavelength dependence of the Verdet constant.[22,23] These glasses, however, are not suitable for refractive elements due to a broad absorption band in the blue spectrum.

The advent of the laser made it necessary to consider also direct interaction between the local electromagnetic field of a high-intensity light beam and the electronic structure of the glass. This effect is described by an intensity dependence of the refractive index,

$$n = n(I) = \sum_{i=0}^{\infty} n_i \cdot I \approx n_0 + gI \tag{15}$$

where n_0 = refractive index at low intensity; I = beam intensity, W/m²; and g = a constant.

Because $I \sim E^2$, Eq. (15) is also written

$$n(E^2) \approx n_0 + n_2 E^2 \tag{16}$$

with E = electric field strength (measured, e.g., in electrostatic units), and n_2 = nonlinear refractive index.

An empirical formula given in Ref. 24, relating n_2 to n_d and v_d, is used to predict n_2 values. For presently available neodymium-doped laser glasses, the values for n_2 are $\approx 10^{-13}$; lithium-alumina-silicate base glasses have the highest values, potassium-baria-phosphate glasses the lowest. A low value is desirable, provided other properties of the glasses are acceptable. At present, nonlinear refractive indexes are important only for glasses used in laser systems.

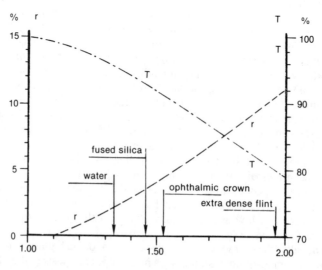

Fig. 3. Reflectivity r (one surface) and transmittance T (two surfaces — entrance and exit planes) for nonabsorbing plate for monochromatic, unpolarized light under normal incidence vs refractive index.

Transmittance

A monochromatic light beam of intensity I_0 entering a planoparallel glass plate is partially reflected, partially absorbed, and partially transmitted. For normal incidence, and in the absence of fluorescence and diffuse scattering, the transmitted intensity I_t is

$$I_t = I_0(1 - r)^2 e^{-kx} \sum_{m=0}^{\infty} \{re^{-kx}\}^{2m} \tag{17}$$

where r = reflectivity = $\{(n - 1)/(n + 1)\}^2$; n = refractive index vs air, $n = n(\lambda)$; k = extinction coefficient = $k(\lambda)$; λ = wavelength, and x = thickness of the plate.

The summation term describes the effect of multiple internal reflections. It converges rapidly to one and can normally be replaced by its limit.

To characterize a glass by the reduction in light transmission from absorption only, the *internal transmittance*

$$\tau_i = 1 - \exp(-kx) \tag{18}$$

is used. This expression is free from reflection losses which, for higher index glasses, can reach noticeable values (Fig. 3). Occasionally the extinction coefficient $k(\lambda)$ or the optical density (kx) are used in place of τ_i.

Because of the low absorptivity of optical glasses, thicknesses of 5 mm or more are frequently required for recognizing characteristic features of the transmission (or absorption) spectrum. Such spectra are very useful for detecting trace impurities in trouble-shooting efforts. Examples of internal transmittance for optical glasses are given in Fig. 4.

Solarization of glasses is particularly detrimental to high-lead glasses used as radiation-shielding windows and in optical systems exposed to high-energy radiation. Although the discoloration fades with time (at room temperature, within several months), it renders the glasses useless for the intended application. To reduce the initial transmission loss caused by high-energy radiation, these glasses are stabilized by addition of CeO_2, elimination of certain fining agents, and redox control during melting.

Chemical Durability

The chemical durability of optical glasses is their most important nonoptical property. No laboratory test simulates completely the different types of attack an optical glass endures during manufacture of the refractive elements and final use. Testing to durability standards for container, plate, or chemical-apparatus glasses would be meaningless, for obvious reasons. International attempts to standardize tests for durability of optical glasses have failed so far. Consequently, various special tests are used throughout the industry: They can be categorized as acid, weathering, and alkali-resistance tests.

A frequently used acid test measures attack by both a strong and a weak acid. Both parts of the test measure the time required to produce, at room temperature, a layer ≈ 100-nm thick at the surface of a polished, flat sample. The layer thickness is estimated by the appearance of thin-film interference colors. A $0.5N$ solution of HNO_3 with pH = 0.3 is used for one test, a standard acetic acid solution with pH = 4.62 for the other. The time required to produce the 100-nm layer by nitric acid attack ranges from >100 h (group 1) to <6 min (group 5). The acetic acid test is applied to rank the very sensitive glasses within group 5. Time requirements range from 10 to 1 h (group 5a) to <6 min (group 5c). Group designation and time requirements vary slightly with manufacturers' specifications.

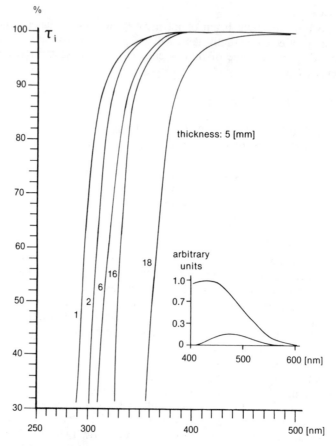

Fig. 4.   Internal spectral transmittance of optical glasses. Curves are numbered according to glass numbers in Tables V and VI. Insert shows fluorescence spectra observed on two optical glasses: Excitation was in the near uv. Similar fluorescence can be observed on some ophthalmic crown lenses. Unless purposely produced, fluorescence spectra and, to a certain degree, also spectral transmittance of optical glasses cannot be considered "typical," as they depend on trace impurities, base-glass composition, and melting conditions.

Another acid test employs two weak acidic solutions. In addition to the acetic acid test already mentioned, testing under attack by a sodium acetate buffer solution of pH = 5.6 is provided for the not-so-durable glasses.

A third acid test was proposed originally to assess the ability of a glass to withstand chemical attack during surfacing operations. Modern washing, grinding, and polishing processes rendered this test less useful for its original purpose, but it remains a meaningful acid-resistance test for optical glasses. In this test polished samples are exposed to $0.5N$ nitric acid. Depending on the duration of the attack and the appearance of the sample's surface after the test, the glasses are rated according to a three-digit, alphanumeric code. The first (numerical) position allows classification in seven groups, based on evidence of attack after specified times. The second (numerical) position rates the glass from "mild attack" (=0.1) to "very heavy attack" (=0.5). The third position defines the appearance of the surface in detail: For instance, the designation 2.2.b means "(signs of attack observed after exposure between 10 and 100 h).(light attack).(uniform removal of material, leaving a polished appearance)."

For comparing the bulk durability of optical glasses, some manufacturers apply powder tests: Glass powder of 420- to 590-μm particle size is suspended in a platinum sieve and boiled in reagent for 60 min. Two reagents are used: distilled water of 6.5 < pH < 7.5 and 0.01N nitric acid. After carefully drying the specimen, its relative weight loss in percent is determined. The water-resistance scale ranges from <0.05% (class 1) to >1.10% (class 6), and the corresponding acid-resistance scale from <0.20% (class 1) to >2.20 (class 6).

For evaluating the resistance of a glass vs atmospheric moisture, also called *weathering resistance,* a freshly polished, flat plate is exposed to air of 85% relative humidity at 50°C for 24 h and subsequently inspected microscopically at 50-times magnification. Depending on the degree of visibility of stain or "dimming" at two specified illumination levels, the glasses are classified into four groups, with group 1 the best.

A more realistic test for resistance against attack by atmospheric moisture exposes polished glass plates to 1-h temperature/humidity cycles between 45° and 55°C for up to 180 h. The degree of light scattering generated by the attack of condensing water vapor is measured on these samples and compared to the scattering observed on three "standard" glasses chosen by the glass manufacturer using the test. The glasses are arranged into four classes: After 180-h exposure, the specimens in class 1 show "no or only slight" surface deterioration; class 4 indicates surface attack clearly observable by light scattering after 30-h exposure. For some very delicate glasses, class 4 is reached after only 5-h exposure.

In modern high-speed cleaning and surfacing operations, optical glasses can be exposed to highly alkaline solutions at elevated temperature. Therefore, an alkali-resistance test introduced by one glass manufacturer was most welcome: In this test, the surface of a glass specimen is exposed to NaOH solution (pH = 10) at 90°C. The time required to destroy a 100-nm-thick surface layer is measured, and the results are arranged into four classes (with class 1 the best) requiring >20 h exposure. As in the third acid test, the specimen appearance after the test serves as an additional means of characterization against alkali attack. A numerical key from 0 (no visible attack) through 4 (a white crust) is used.

Opportunities for increasing the chemical durability of some optical glasses still exist. Because of the complexity of compositions, very few general rules can be applied, and all of them are known to the glass technologist. For instance, glasses with relatively high amounts of low-water-soluble oxides (SiO_2, Al_2O_3, TiO_2, rare earths) usually have high acid resistance and are durable to climatic variations. Glasses containing large amounts of alkali oxides (P_2O_5, B_2O_3, or fluorides) are usually of low chemical durability. There are, however, exceptions, as shown by some zinc-alumina-phosphate glasses. The challenge lies in using mutual interaction of glass constituents or addition of small amounts of compounds that are effective in specific compositional systems. General guidelines, although not readily accessible, can be found in the theories presented by Weyl and Marboe.[25]

Thermal Expansion

The most widely used glasses have coefficients of linear thermal expansion

Table IV. Correlations between Density (ρ) and Refractive Index for Selected Glass Groups

Glass Type	a	b	Index Range	Density Range	Sample Size Used	R^2
Silicates	1.29	0.38	1.488 to 1.953	2.37 to 6.26	195	0.96
Titania-borosilicates, >10% TiO_2	1.43	0.43	1.590 to 1.850	2.60 to 4.53	7	0.99
Borates, >25% B_2O_3	1.38	0.39	1.530 to 1.961	2.52 to 6.13	34	0.95
Fluoro-borosilicates, >10% fluorides	0.14	0.87	1.446 to 1.617	2.27 to 2.79	7	0.91

The formula $n^2 = a + b\rho$ is useful for estimating refractive indexes from density measurements. R^2 is the coefficient of determination and expresses the goodness of fit of the curve to the data, with 1.00 best. The column "sample size" lists the number of glasses of different composition on which calculation of the coefficients a and b is based.

of $\approx 90 \times 10^{-7}\,°C^{-1}$, low-expansion borosilicate glasses of $\approx 30 \times 10^{-7}$ to $50 \times 10^{-7}\,°C^{-1}$; optical glasses range from $\approx 35 \times 10^{-7}$ to $150 \times 10^{-7}\,°C^{-1}$.

Knowledge of the thermal expansion of optical glasses is particularly important in selecting process parameters for surfacing operations. Glasses with high thermal expansion, such as, for instance, some fluor-crown or low-flint glasses, should not be exposed to temperature shock, a precaution unnecessary for glasses of medium and lower expansion. Low-expansion glasses are not only easier to surface, but also are used extensively as mirror blanks for smaller telescopes with reflective optics. Today, such mirrors are mostly fused silica or zero-expansion, transparent glass-ceramic. It is interesting to note that the (nonoptical) glass company that introduced into trade the first transparent, zero-expansion glass-ceramic now offers it, although under a different trade name, for another advanced optical device, the laser gyroscope.

Thermal expansion of an optical glass becomes important for predicting optical behavior of lens systems which are to be exposed to temperature variations. Combined with the temperature change of the refractive indexes of a glass, discussed previously, thermal expansion makes it possible to design and build lens systems with almost constant power over a given temperature range. The same two properties are also used in "athermalized" laser glasses. Glass-property requirements for athermalizing an optical system depend on details of its design. For instance, using the notation in Eqs. (2) and (14), an athermalized, thin, single lens requires a glass meeting the condition

$$n_{g_0}(\alpha - b_g) = n_{a_0}(\alpha - b_a) \tag{19}$$

with α = coefficient of linear thermal expansion of the glass. Even for the thin lens, the athermalization condition is wavelength-dependent. Therefore, a "thermo-optical" or "opto-thermal" coefficient of the first or second order, composed of the coefficient of expansion and the temperature coefficient of a single refractive index, should be used with caution.

Other Physical Properties

Density values for optical glasses range from ≈ 2.4 to 6.3 g/cm^3. Since optical glasses are sold by weight, the density must be known for each glass. This readily available property can also be used for estimating refractive indexes. To find, for instance, glasses with a high index-to-weight ratio ("light-weight" ophthalmic crowns), one would select compositions with high values of the constants a and b in Table IV.

Elastic properties, viscosity data, thermal conductivity, and other physical properties of optical glasses are needed occasionally for special applications. Viscosity vs temperature functions, for instance, became important during the early phases of fiber-optics development. Due to the wide compositional ranges encompassed by optical glasses (alkali-lime silicates, lead silicates, borosilicates, rare-earth silicates and borates, silicophosphates, etc.) their viscosity curves represent those of conventional, as well as very short and very long, glasses. For the same reason, one can find a relatively wide range of values for Young's modulus, namely ≈ 48 to 82 MPa (7 to 12 Mpsi). Poisson's ratio varies from ≈ 0.19 to 0.31. Thermal and elastic aftereffects are important for estimating the long-term behavior of large glass parts in special systems; some data are given by Deeg.[10] Knoop and Vickers hardness measurements show that the bulk hardness of fresh fracture surfaces is higher than that of a polished surface. This result is important first for selecting proper surfacing conditions and then for predicting probable reactions to abuse of a polished lens surface. Among thermal properties specific heat, thermal conductivity, and thermal diffusivity are required occasionally.

Manufacturing Aspects

Examples of Glass Compositions

The key to success in making available a suitable selection of optical glasses was the introduction of elements not used previously as glass ingredients. Relationships between glass composition and optical properties have been studied thoroughly. The results of these efforts with potential for industrial application are found primarily in the extensive, still-growing patent literature. To demonstrate the variety of glass compositions for optical applications, a few examples are given in Tables V and VI. These examples by no means represent the entire compositional range of the more than 200 optical glasses presently on the market: For instance, no thorium oxide-containing glasses are listed, although they were quite popular for many years as supplements to and substitutes for lanthanum oxide-containing glasses. A study of the patent literature of the past 30 years is highly recommended for information on recent optical-glass compositions.

Melting, Forming, and Annealing

Production of high-quality optical glass begins with proper selection of raw materials and a correctly balanced batch. In addition to chemical composition and raw-material purity — in certain cases down to the ppb range — the particle size distribution of the mixed batch is important. Partial segregation during transport can later cause striae, thus reducing yield. Optical-glass companies usually control their sand mines and acid-washing plants; any incoming raw material is thoroughly analyzed. Each manufacturer has his own purity specifications and justifiably does not publish them in detail.

Optical-glass melting uses platinum crucibles; occasionally, still, large clay

Table V. Examples of Compositions of Optical Crown Glasses, in Round Numbers

	Representative for Glass Type							
	PK	BK	PSK	K	BaK	SK	SSK	LaK
					Glass Number			
Ingredients	1	2	3	4	5	6	7	8
SiO_2	68	72	55	75	48	41	37	6
ZrO_2								6
B_2O_3	14	12	12		4	5	6	40
Al_2O_3			2	1	1	2	2	
La_2O_3								30
CaO				4				17
BaO	1	1	22		29	42	40	
ZnO		2			9	9	8	
PbO							4	
Na_2O	8	4	5	9	1			
K_2O	8	8	4	11	7			

Additional oxides found in crowns include P_2O_5, CeO_2, TiO_2, MgO, SrO, Li_2O, and the fining agents Sb_2O_3 and As_2O_3. Small amounts of halides such as CaF_2, KHF_2, NaCl, KBr, and NH_4Cl are occasionally included in the batches. An example of a glass of the type FK is (in round at.%) 23.0Si-6.0B-0.2Al-23.8K-2.1F-44.8O with $n_D = 1.4785$ and $\nu_D = 70.2$. The above identified glasses are shown in the $\{n_d; v_d\}$ diagram of Fig. 2.

Table VI. Examples of Compositions of Optical Flint Glasses, in Round Numbers

	Representative of Glass Type									
	KF	BaLF	LLF	BaF	LaF	LaSF	LF	F	BaSF	SF
					Glass Number					
Ingredients	9	10	11	12	13	14	15	16	17	18
WO_3						7				
Ta_2O_5						8				
SiO_2	67	54	61	56	4	6	53	44	42	27
ZrO_2					6	6				
B_2O_3					32	17				
La_2O_3					19	24				
CaO					14					
BaO		14		12	8	11			11	
ZnO	3	10			12	4			5	
CdO						14				
PbO	13	11	26	17	6		34	46	34	71
Na_2O	16	2	5	2			4	2	1	1
K_2O		9	8	13			7	7	7	1

Additional oxides found in optical flints include P_2O_5, Nb_2O_5, TiO_2, Al_2O_3, Sb_2O_3, and Li_2O. The same compounds listed in Table V are used as fining and flux additions.

pots holding about one ton of glass, continuous tanks, and numerous melters of truly unique design are used. The variety of melting units is caused by the range of glass and batch compositions, the different liquidus points (some optical glasses have more than one) and degrees of devitrification tendency, the effect of melting

Table VII. Effect on Refractive Index of Mutual Substitution of Small Amounts ($\approx$1 wt%) of Glass Oxides

Replacement by	Oxides to be Replaced														
	Li	Na	K	Mg	Ca	Ba	Zn	Cd	B	Al	La	Si	Ti	Zr	Pb
Li	o	–	–	–	d	–	–	d	–	–	d	–	D	d	d
Na	d	o	–	d	D	d	d	D	d	–	D	–	D	D	D
K	D	d	o	d	D	d	D	D	d	d	D	–	D	D	D
Mg	d	–	–	o	d	d	d	d	–	–	d	–	D	d	d
Ca	d	–	–	–	o	–	–	d	–	–	d	–	D	d	d
Ba	d	–	–	–	d	o	d	d	–	–	d	–	D	d	d
Zn	d	–	–	–	d	–	o	d	–	–	d	–	D	d	d
Cd	d	–	–	–	d	–	–	o	d	–	d	–	D	D	d
B	d	d	–	d	D	d	d	d	o	–	D	–	D	d	D
Al	D	d	d	d	D	d	d	D	d	o	D	–	D	D	D
La	d	d	–	d	–	–	–	d	–	–	o	–	D	d	d
Si	D	–	d	d	d	d	d	–	d	d	O	o	O	–	D
Ti	–	–	–	–	–	–	–	–	–	–	–	–	o	o	–
Zr	–	–	–	–	–	–	–	–	–	–	–	–	d	o	d
Pb	–	–	–	–	–	–	–	–	–	–	–	–	D	d	o

The oxide symbol is replaced by the corresponding atomic symbol. Zn, Cd, Pb represent the monoxides; Si and Ti the dioxides; B and Al the oxides B_2O_3 and Al_2O_3, respectively. An i or l means increase and d or D decrease, with capital letters indicating index changes >0.003. The information presented here is used to correct for minor production variations, mostly in continuous operations.

atmosphere on glass properties, the different attack rate of the glasses on refractories, variable market demand, and so forth.

While a melt is in progress, refractive index samples are taken and, if required, batch corrections are made. Tables such as Table VII are used in these efforts. Instead of the semiquantitative information given there, they contain rather precise data incorporating specific process parameters. Some melts must be carried out under a controlled atmosphere, in electrical, direct- or induction-heated furnaces. Ultraviolet-transmitting glasses require extremely pure raw materials, carefully balanced batches, and atmosphere control that may have to follow the batch reactions. Selecting the right fining agent, e.g., As_2O_5 instead of As_2O_3, can mean the difference between success and failure. Infrared-transmitting glasses are made, sometimes, from presintered batch melted in a dry atmosphere or in vacuo, whereby a change in furnace atmosphere during the melt may take place. High-volume optical glasses are produced in continuous tanks of an approximate throughput of one ton/day or less. Fining (forehearth) and feeding sections of the tanks are platinum-lined and equipped with platinum stirrers. The melter can be the more conventional rectangular type, combining fossil fuel and electric booster heat, or of the more advanced octagonal or hexagonal shape, with all-electric heat. The most widely used electrode material is probably tin oxide; occasionally, water-cooled molybdenum or platinum electrodes are still found. Selecting the right refractories for crucibles, pots, and tanks is particularly important, and optical-glass manufacturers maintain in-house expertise in this area. The many ceramic factories that produced large melting pots within a glass plant 25 years ago have disappeared. Details on melters suitable or adaptable for optical-glass production are available in the patent literature. Designs of the best versions or the optimal mode of operation of a given melter for the various glass types, however, are difficult to find even there. A small melter that can be equipped with platinum or ceramic crucibles and produce high-quality glass at high yield (75–80%) is described by Deeg and Silverberg.[26]

Glass-forming processes and equipment vary from plant to plant. Both are chosen as required by the form of supply offered. Casting tables are still used, although most optical-glass items are formed on relatively small, slow presses, also used for preparing glass gobs. Optical and even ophthalmic glass presses are designed, or, as in other glass companies, at least specified by the glass manufacturer's technical staff. Strips and rods are extruded or continuously cast in inclined channels. Pressings for small orders are made by hand from cut and tumbled blanks of defined volume. Again, for details on optical glass forming, the patent literature should be consulted and critically evaluated. In some cases, this will require experimental work to determine compatibility of a new forming mechanism with individual glasses and already-installed melters and annealers. Various steels, including stainless versions and, in special cases, beryllium bronze or graphite-based materials, are used as mold materials.

Annealing mass-produced small items (pressings or gobs) takes place in continuous lehrs, fine ("special") annealing usually in air-circulating electric furnaces. Pieces weighing several hundred kilograms ("massive optics") are fine-annealed in individual bell furnaces with electric top, side, and bottom heat. In each case precise temperature control and strict adherence to preestablished annealing schedules are exercised.

Quality Requirements and Inspection

Quality assurance (QA) departments in optical-glass plants are precision laboratories well equipped with optical and electron microscopes, spectrophotometers, prism spectrometers, interferometers, and various shadowgraphs. They are supported by analytical and nonoptical metrological laboratories.

The most important properties of an optical glass are its refractive indexes: Considerable effort therefore is spent in measuring and controlling them. Instrumentation used routinely for production control permits measurement of n at a given wavelength with an accuracy of $\pm 3 \times 10^{-5}$ and, for special requirements, $\pm 5 \times 10^{-6}$. Systematic errors usually are eliminated by applying two independent methods. The index values given in the data sheets of optical glasses represent mean values of many melts. Thus, each individual melt can deviate in its n_d value from the data-sheet value by $\pm 1 \times 10^{-3}\%$ and, for higher index glasses, twice that. It is, therefore, accepted practice that each shipment of optical glass be accompanied by a "melt sheet," certifying actual values of important index and dispersion data.

Refractive indexes are measured most frequently with prism spectrometers, requiring not only skilled laboratory technicians but also access to an outstanding optical shop for preparing the test prisms. The light sources are gas-discharge lamps with well-defined monochromatic emission lines (Table II); double lines — for instance, the sodium D line — do not permit index measurements at the level of precision required today.

As do other properties, the refractive index of a glass of given composition depends on the thermal history of the sample. This fact is used for index adjustment of coarse, annealed optical glasses. Slow annealing rates (below $\approx 5°C/h$) throughout the entire volume of the glass piece can increase the refractive index into the fourth or even the third decimal place. This rule assumes an "average" or "customary" prior thermal history of the article to be annealed. Precision annealing schedules have been developed not only for each glass but also for different-volume articles of the same glass. With annealing rates in the order of a few degrees per hour, and considering the low thermal diffusivity of glass, it is not surprising that massive pieces of optical glass are annealed for several months.

The optical glassmaker distinguishes between "striae" and "optical homogeneity." Striae are highly localized, cordlike variations of the refractive index. Optical homogeneity measures a smooth, gradual change in refractive index throughout an article. Its degree is measured interferometrically and, within a given melt, is usually better than a corresponding index variation of $\pm 1 \times 10^{-4}$ or, for specially selected glass pieces, $\pm 1 \times 10^{-6}$. Striae are detected and rated by shadowgraphs, using high-intensity point-light sources. For sorting out glass of the highest quality, the more sensitive Schlieren methods, for instance knife-edge tests, are used. Under such rigorous inspection techniques, many optical glasses are not striae-free, and the data sheets contain information on type and expected intensity of striation. Efforts to improve striae quality by modifying melting techniques and batch compositions are carried out continuously. In many cases, striae are caused by selective volatilization of glass ingredients, which has led to the introduction of special crucible and furnace designs.

At the precision level required for optical glasses, even the best-annealed glass can show measurable degrees of birefringence. The leading optical-glass suppliers publish limits of birefringence for their regular production items: that is, for glass types, shapes, and sizes of blanks. High birefringence values for regular production items are approximately 20 nm/cm. Special annealing can reduce this level to less than 4 nm/cm (about 1/100 the wavelength of blue light per centimeter!), depending on glass type and blank size.

Striae, birefringence, or optical inhomogeneity cannot be eliminated entirely. The same principle applies to solid or gaseous inclusions. These inclusions are

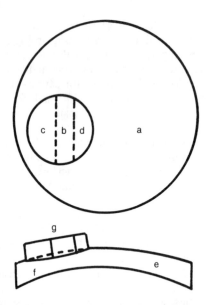

Fig. 5. Fused trifocal ophthalmic lens: (*a*) distance-vision portion, (*b*) intermediate-vision portion, (*c*) near-vision portion, (*d*) (invisible) "upper" segment portion. Notation differs slightly if applied to the fused blank prior to surfacing: (*e*) major, (*f*) countersink, (*g*) (segment) "button." Prior to fusion, lower segment surface, as well as countersink, are ground and polished to defined curves, with radius of curvature of countersink slightly larger than that of segment. Segment itself consists of portions fused together, as indicated.

treated, for optical purposes, as one kind of defect, called *bubbles*. Most optical glasses are bubble-free. For extremely difficult and relatively new glasses, where the manufacturing processes have not yet been perfected, bubble group classifications are introduced and the glasses are ranked accordingly. The classification scheme addresses inclusion size, number of bubbles allowed in an article of given size, and total projected area of inclusions per volume. The "total bubble area" is the sum of the cross sections of all individual inclusions. For a very good glass the total bubble area is <0.03 mm^2/100 cm^3; a poor optical glass has a total inclusion area >0.5 mm^2/100 cm^3. All bubbles >50 μm in diameter are included in classifications of regular optical glasses. For special applications the selection is even tighter, including, for instance, bubbles <50 μm in diameter.

Ophthalmic Glasses

Index and dispersion requirements for ophthalmic products are far less stringent than for optical instruments. The problem of achromatization has been addressed in connection with fused multifocal lenses (Fig. 5), but turned out to be less of an issue than finding durable, high-index, "segment glasses" that could be fused to a carrier blank ("major") without stress and without distorting and opacifying the optically defined fusion interface between them. Although index specifications cover only the third decimal point for ophthalmic crown glasses and the fourth for ophthalmic segments, striae and inclusions that could interfere

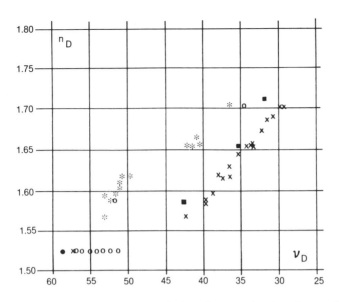

Fig. 6. Representation of ophthalmic glasses in the $\{n_D; \nu_D\}$ plane: × = clear crown and flint glasses, * = clear barium crowns and flints, ○ = tinted glasses, ● = photochromics, ■ = low-expansion flints (fusible to borosilicate-based photochromics). Terms "crown" and "flint" applied to ophthalmic glasses are not used in exact sense as for optical glasses. Glasses used for single-vision lenses and for major (usually distant-vision or "upper" part) of fused multifocal lenses have a refractive index of 1.5231 and, depending on special features, Abbe numbers from ≈51–59.

with normal vision are not acceptable. Ophthalmic glass lenses are among the few consumer products that meet scientifically established, exceptionally strict requirements.

Today, high-quality spectacle lenses are mass-produced from glass blanks, frequently on automated, computer-controlled production lines. This process requires uniform critical properties in the glass blanks delivered to the lens manufacturer. These critical properties include dimensions; dimensional tolerances and surface texture of the glass pressings; degree of annealing and the glass properties n_D, ν_D; chemical durability; density; strain, annealing, and softening points; coefficients of linear thermal expansion and contraction; and color (defined either in terms of spectral transmissivity or tri-stimulus values).

The near- and distant-vision portions of a fused multifocal lens share the same front surface curve. Since the near-vision portion requires a higher power, it must be made of a glass with a refractive index higher than that of the distant-vision portion. To assist the lens manufacturer in producing different "add" powers without increasing the number of front and countersink curves, the glass industry offers a variety of high-index glasses. Figure 6 illustrates the location of some ophthalmic glasses in a $\{n_D; \nu_D\}$ diagram, and examples of nominal compositions are given in Table VIII.

The need for eye protection against radiation, the discovery of photochromic glasses, and fashion trends brought about a wide spectrum of colored ophthalmic

Table VIII. Nominal Compositions and Approximate Refractive Index n_D for Some Ophthalmic Glasses

	19	20	21	22	23	24	25	26	27	28
	Ophthalmic Crowns					Ophthalmic Flints				
P_2O_5					20					
SiO_2	67	68	70	56	21	47	48	45	39	34
TiO_2	t	t	1		t	3	4	1	4	7
ZrO_2				2	2	2		1	6	5
B_2O_3				16	12	5	4		2	6
Al_2O_3	2	2		9	22	1		1		
MgO	t				t					
CaO	9	9	6		3	4	6		6	8
SrO							t			t
BaO				7	9	22	13		19	14
ZnO	2	3	5			2				
PbO				6			14	40	16	18
Li_2O				3	t					
Na_2O	7	8	9	2	2	6	7	3	6	8
K_2O	11	9	8		7	7	2	8	1	t
n_D			1.523			1.59	1.60	1.62	1.66	1.68

Concentrations below 1 wt% are indicated by t; fining agents and fluxes are not included.

glasses. They all meet refractive index requirements, as indicated in Fig. 6, as well as color specifications. Desirable but not yet available are glasses for preparing one-piece, polarizing, photochromic lenses.

Glasses for Active Optical Elements

In a very broad sense, active optical elements may be defined as glasses affecting a beam of light beyond merely changing its direction. This definition includes filter and laser glasses, glasses for Faraday rotators and Cerenkov counters, and photochromic glasses. The most widely used and commerically important optical elements are the filter glasses. Laser glasses have limited application among the other laser materials. Faraday rotating glasses have been discussed briefly, and glasses for Cerenkov counters are usually selected from existing optical glasses. Photochromic glasses, which are particularly important as ophthalmic products, are discussed elsewhere.[27] Only the two first groups, therefore, will be discussed briefly.

Filter glasses are primarily used in the form of flat plates positioned normal to the optical axis of an optical system. Their refractive power is therefore of only minor importance. Instead, special emphasis is placed on internal spectral transmittance and chromaticity coordinates. A typical data sheet for a high-quality filter glass includes the following information.

1. Internal transmittance in tabular and graphic form.
2. Tri-stimulus values for at least one, usually two, standard illuminants.
3. Tolerances of internal transmittance data for selected wavelengths.
4. Refractive indexes for selected wavelengths, up to three decimal points.
5. Qualitative information on fluorescence.

6. Remarks on special behavior (solarization, precautions to be taken for striking glasses).

7. Density.

8. Coefficients of linear thermal expansion for two temperature ranges.

9. Chemical durability.

Graphic representation of the internal spectral transmittance of optical filter glasses uses the expression

$$\theta = 1 - \log[\log(1/\tau_i)] \tag{20}$$

where τ_i = internal spectral transmittance. For a given sample thickness, $\theta(\lambda)$ is plotted in a grid with a double logarithmic scale at the ordinate (θ) and a linear scale at the abscissa (λ). A linear translation of the $\theta(\lambda)$ curve parallel to the θ axis converts internal transmittance values to other glass thicknesses.

$$\theta_1 = \theta_2 - \log(t_1/t_2) \tag{21}$$

where the index "1" denotes transmittance, θ, and, t, thickness, respectively, for one sample, the index "2" for the other.

Optical filter glasses can be divided into four groups:

1. Transmitting in the uv or ir, but absorbing in the visible (black or very dark glasses).

2. Transmitting in the ir and the visible, but absorbing in the uv (clear or "white" glasses, occasionally light yellow).

3. Transmitting in the visible, but absorbing in the ir ("heat-screen" glasses).

4. Selectively absorbing in the visible (colored glasses).

Representative transmission curves of optical filter glasses in the conventional τ_i, not the θ_i, format are shown in Fig. 7. Some colorant information is listed in Table IX.

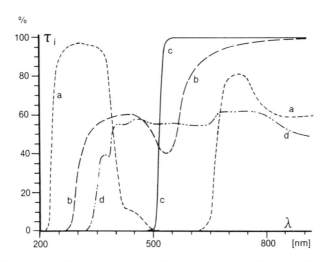

Fig. 7. Examples of spectral internal transmittance of optical filter glasses: 1-mm sample thickness. Colorants for different glasses are (a) Co_2O_3 and Ni_2O_3, (b) $AuCl_3$ (striking glass, "gold ruby"), (c) CdS and Na_2SeO_3 (striking glass, "selenium ruby"), and (d) Fe_3O_4 and Co_2O_3.

31

Table IX. Colorants and Selected Base Glasses for Optical Filter Glasses

Filter Characteristics	Colorants	Base Glasses
uv transmitting, black	Co_2O_3 and Ni_2O_3	alkali-lime silicate, zinc silicate, lead borosilicate
uv transmitting, violet	$KMnO_4$ or Co_2O_3	lead silicate, alkali-lime silicate
Blue	Co_2O_3 and CuO Co_2O_3, CuO, and $KMnO_4$	alkali-zinc silicate, alkali-lead silicate
Blue-green	CuO	alkali-alumina phosphate, alkali-zinc silicate
Yellow, low fluorescence	$K_2Cr_2O_7$	alkali-lime silicate
Yellow, orange, red (striking glasses)	CdS and Na_2SeO_3	alkali-lime silicate, alkali-zinc silicate
Pink (striking glasses)	$AuCl_3$	antimony borosilicate
Purple	Nd_2O_3, Didymiumoxide	alkali-lime silicate, alkali-zinc silicate, phosphate-based
Brown (striking glasses)	$AuCl_3$ and Fe_2O_3	antimony borosilicate
Neutral (gray)	Fe_2O_3 and Co_2O_3	borosilicate
ir transmitting, red	As_2S_3 and other chalcide glasses	
ir transmitting, black	MnO_2 or MnO_2 and $K_2Cr_2O_7$	alkali-zinc silicate, alkali-lead silicate
Heat-absorbing, greenish-blue tint	Fe or FeO and Zn or Fe and Fe_3O_4	phosphate-based, alkali-zinc silicate
uv absorbing, yellowish	CeO_2	alkali-lead silicate

Batch calculations follow the rule to add the colorants to 100 parts (wt) of the base glass, calculated in oxides. Actual colors are influenced, in some cases even determined, by the melting conditions and minor batch ingredients such as NH_4OH, NaCl, or SnO, etc.

Optical filter production uses the same basic coloring mechanisms as does the container, tableware, and art-glass industry. Pigments, however, are not used. Emphasis is on the reproducibility of the colors or tints, which requires well-defined raw materials, including the coloring agents, well-balanced batches, and, in many cases, close control of the redox conditions in and above the melt. For temperature-colored glasses (striking glasses, "ruby" glasses), the subsequent heat treatment must be controlled thoroughly, in addition to the other process parameters mentioned. Computer-controlled melting and annealing, combined with X-ray fluorescence analysis, play important roles in continuous production.

In the most frequently manufactured laser glasses, Nd^{3+} is the active ion. About 20 years of development produced a set of properties describing sufficiently the few laser glasses available in commerce; glass selection for specific applications is based on these data. Some of the laser properties are determined by the base glass, which is, therefore, identified in the designation (silicate or phosphate glasses). The neodymium concentration is reported either in terms of number of Nd^{3+} ions per unit volume or weight percent Nd_2O_3. The field is still developing, as the laser-glass producers offer to melt glasses with neodymium concentrations specified by the customer. Among properties determining the laser action are emission wavelength (≈ 1060 nm for silicate-based and ≈ 1050 nm for phosphate-based glasses), width of the emission line (≈ 20 nm for phosphate and 30 nm for silicate glasses), fluorescence lifetime (> 300 μs), estimated damage threshold (usually > 10 J/cm^2 for silicate- and 25 J/cm^2 for phosphate-based glasses, but depending on pulse lengths), and induced-emission cross section (≈ 3–4×10^{-20} cm^2). The induced-emission cross section, with the volume concentration of pumped Nd ions, determines the gain coefficient of the laser rod. In most cases, the laser-property measurements are not yet as precise as the optical-property measurements described previously. Variances of several percent, depending on the method used, are common. Among optical properties listed are refractive indexes for at least the laser-emission wavelength, the "nonlinear refractive index," Brewster's angle for the emission wavelength, and other properties described previously. Importantly, chemtempering, a process that did not originate in the optical-glass industry, is offered for laser-glass rods.

Summary

Historically the first "high-tech materials," optical glasses are produced in a wide variety of compositions, to exact optical specifications. Their properties and quality are well defined in extensive data sheets. Manufacturing requires chemically well-defined raw materials of high purity and careful control of batching, melting, and annealing parameters in units that are frequently dedicated to specific compositional groups. Quality control and assurance uses advanced optical instrumentation.

References

[1]H. Hovestadt, Jena Glass and Its Scientific and Industrial Applications. Translation of the original German edition of 1900 by J. D. and A. Everett. (The English edition also contains several amendments updating the original version). MacMillan, London, 1902.

[2]E. Zschimmer, Die Glasindustrie in Jena. Eugen Diederichs Verlag, Jena, 1912.

[3]E. Zschimmer, Theorie der Glasschmelzkunst als Physikalisch-Chemische Technik, Vol. I. Verlag Volksbuchhandlung GmbH, Jena, 1911.

[4]H. Thiene, Glas, Vol. I. Gustav Fischer Verlag, Jena, 1931.

[5]R. Guenther, Glass-Melting Tank Furnaces. Soc. Glass Technol., Sheffield, 1958.

[6]W. Kiaulehn, Der Zug der 41 Glasmacher. Jenaer Glaswerk Schott & Gen., Mainz, 1959.

[7]E. Schott (editor), Beitraege zur angewandten Glasforschung. Wissenschaftl. Verlagsges. m.b.H., Stuttgart, 1959.

[8]T. C. Barker, Pilkington Brothers and the Glass Industry. George Allen & Unwin Ltd., London, 1960.

[9]G. Hetherington, Portrait of a Company. TSL Thermal Syndicate p.l.c., Wallsend, 1981.

[10]E. Deeg, "Optical Glasses," Proceedings of the VIIth International Glass Congress, Brussels, 1965. Paper I.1.–I.11.

[11]W. V. Harcourt, "Report on a Gas Furnace for Experiments on Vitrification and Other Applications of High Heat in the Laboratory," Brit. Assoc. Rept. of the XIV meeting, York, 1844.

[12]G. G. Stokes, "Notice of the Researches of the Late Rev. William Vernon Harcourt, on the Conditions of Transparency in Glass and the Connexion Between the Chemical Constitution and Optical Properties of Different Glasses," Brit. Assoc. Rept. of the XLI meeting, 1871.

[13]E. Abbe, Die optischen Hilfsmittel der Mikroskopie, Bericht ueber die wissensch. App. a. d. Londoner intern. Ausst. i. J. 1876, I. p. 417. Braunschweig, 1873. Quotation after Ref. 1, p. 3.

[14]G. W. Morey, U. S. Pat. No. 2 150 694, 1937.

[15]K. H. Sun, U. S. Pat. Nos. 2 466 507 and 2 466 509, 1949.

[16] S. Czapski and O. Eppenstein, Grundzuege der Theorie der optischen Instrumente nach Abbe. J. A. Barth, Leipzig, 1924.

[17] M. Herzberger, Strahlenoptik. Springer-Verlag, Berlin, 1931.

[18] A. E. Conrady, Applied Optics and Optical Design. Dover Publications, New York, 1957.

[19] M. Born and E. Wolf, Principles of Optics, 2d ed; Ch. 4.7. Pergamon, Oxford, 1964.

[20] OSA Handbook of Optics. McGraw-Hill, New York, 1978.

[21] F. Reitmayer and E. Deeg, "Optical and Mechanical Properties of Bubble Chamber Windows, *Appl. Optics,* **2**, 999–1002 (1963).

[22] A. B. Villaverde and E. C. C. Vasconcellos, "Magnetooptical Dispersion of Hoya Glasses AOT-5, AOT-44B, and FR-5," *Appl. Optics,* **21**, 1347–48 (1982).

[23] J. A. Davis and R. M. Bunch, "Temperature Dependence of the Faraday Rotation of Hoya FR-5 Glass," *Appl. Optics,* **23**, 633–36 (1984).

[24] M. L. Boling, A. J. Glass, and A. Owyoung, "Empirical Relationships for Predicting Nonlinear Refractive Index Changes in Optical Solids," *IEEE J. Quantum Electron.* **QE-14**, 601 (1978).

[25] W. A. Weyl and E. C. Marboe, The Constitution of Glasses, Vols. I–III. Interscience, New York, 1962–1967.

[26] E. W. Deeg and C. G. Silverberg, "Melter for Optical Glasses," *Glass Ind.,* April 1970, pp. 164–67.

[27] R. J. Araujo, "Photosensitive Glasses"; this proceedings.

Trade literature issued by the following companies was consulted:

Chance-Pilkington, Lathom, UK
Corning France, Avon, France
Corning Glass Works, Corning, NY
Heraeus Amersil, Inc., Sayreville, NJ
Hoya Glass Works, Ltd., Tokyo, Japan
Jenaer Glaswerk Schott & Gen. Mainz, FRG
Mashpriborintorg, Moscow, USSR
Ohara Optical Glass, Inc., Sagamihara, Japan
Owens-Illinois, Toledo, OH
Schott Optical Glass, Inc., Duryea, PA
Thermal American Fused Quartz Co., Montville, NJ

Additional information on optical glasses can be found in:

W. Eitel, M. Pirani, and K. Scheel, Glastechnische Tabellen. Springer-Verlag, Berlin, 1932 (lithoprinted by Edwards Brothers, Inc., Ann Arbor, MI, 1944). This 700+ page book represents a thoroughly edited, extensive tabulation of glass properties and compositions.
E. Wolf (editor), Progress in Optics, Vols. I–XIX. North-Holland, Amsterdam, New York, Oxford, 1961–1981. This series contains advanced reviews of optical topics, including materials.
MIL-G-174A, Nov. 5, 1963 (with amendment 2, June 25, 1974).
MIL-O-13830A, Sept. 11, 1963 (with amendment 2, July 9, 1982). Both specifications are of interest for quality definitions of optical glasses and finished optical components.
D. F. Horne, Optical Production Technology. Crane, Russak & Co., Inc., New York, 1972. This book covers primarily optical surfacing techniques, including equipment, abrasives, polishing pads and compounds, inspection.
I. Vanderlik, Optical Properties of Glass. Elsevier, New York, 1983.

Container Glass

R. J. RYDER

Brockway Glass Containers
Brockway, PA 15824

J. P. POOLE

Consultant
Brockway, PA 15824

This paper attempts to review the development of the glass-container industry in this country from just before World War II to the present. Since the industry itself can be considered to be 2000 years old, the technical progress made during only the past 40 years is truly remarkable. In spite of these past achievements, however, the glass-container industry today faces its greatest challenge: Can it survive among the host of competitive packaging materials that are now arrayed against it?[1]

Historical Perspective

Before one can appreciate the technical progress the glass-container industry has made since World War II, we must describe the industry as it was about 1940. First, however, note that the glass-container industry was the one industry that came through the depression of the 1930s with full employment, full operating capacity, and a profit—the textbook example of a stable growth industry.

Although considerable, excellent work had been done by such early researchers as W. E. S. Turner and his students at Sheffield, Howard Lillie at Corning, George Morey at the Geophysical Laboratory, Donald Sharp at Hartford-Empire, and G. Tamman of Germany, to name but a few, glassmaking in 1940 was still an art, not a science. The causes of most production upsets were never understood: Changes in apparent workability, sudden increase in seeds, or the occurrence of stones or blisters were unpredictable and always of mysterious origin.

For example, in at least one plant, the color mix to make amber glass was known only to the plant manager, who composed it secretly at 4:00 A.M. every day. Glass temperatures were mostly determined by eye. Optical pyrometers were rare. Furnace-combustion and producer-gas analyses were made with Orsat equipment and took most of the day. Density had been found to indicate a change in glass composition, and some plants measured it daily, using the Archimedes method.[2] To check annealing, nails were often put into bottles and the bottles shaken; if they didn't break, the annealing was assumed adequate. The cause of cords was finally fully explained by M. Knight.[3] Glass tanks typically lasted one year, being put down for a rebuild every Christmas.

In 1940, the container industry produced 54 million gross of bottles, the largest volume in its history. The total sales value was $175 million.[4] At that time, medicine and toiletries, food, and liquor each accounted for about 25% of total glass output. Soft drinks and beer together accounted for only 11%, although 91% of all packaged beer was in glass. But, as the first harbinger of the future, milk

35

began to be packed in paper cartons, and no-deposit bottles — which some people thought were made from soybeans, and were not really glass — were introduced. On December 11, 1939, a suit was filed against 12 corporations including Hartford-Empire, Corning, GCA, and some 100 individuals for violating the Sherman-Clayton Antitrust Act.[5] The outcome, several years later, changed forever the entire structure of the glass-container industry.

Various forming machines were used. The first truly automatic forming machine was developed in 1904 by M. J. Owens,[6] and many Owens machines were still operating in 1940. In addition, various types of Lynch 10s, Miller press and blows, and a few Hartford I.S. machines were operating.[7] Sorting, inspection, and packing were all done by hand. Some lehrs were still loaded manually.

After World War II, the glass-container industry was optimistic about solving its many practical operating problems through improvements in technology gained by its ever-increasing R&D efforts. This attitude prevailed through the 1970s when, in spite of many major technological successes, the industry began to lose more and more of its market to competitive packaging.

In 1984, the prevailing mood in the glass-container industry was pessimism, and the industry is now struggling for survival by heroically increasing productivity and decreasing costs. However, this has also forced personnel layoffs, plant closings, price cutting, and, unfortunately, drastic curtailment or elimination of R&D spending.

Raw Materials

Disregarding cost, the major changes in raw materials over the past 40 years were in particle-size specifications and the number of minor ingredients used in the glass batches. Many of the materials such as sand, limestone, and cullet are less pure now than they were then.

A typical flint-container batch in the early 1940s contained sand; limestone and/or dolomite; feldspar, aplite, or nepheline syenite; and soda ash, barytes, saltcake, fluorspar, arsenic, nitre, selenium, cobalt, and borax. Particle sizes were specified in a general way, particularly for the major ingredients. Coarser grain fractions were permitted then than now. Because of cost, a typical flint-container batch today consists of only sand; feldspar, aplite, or nepheline syenite; limestone, soda ash, selenium, and saltcake; and carbon or slag — a total of 7 ingredients now, compared to 13 then. Grain-size specifications are very restrictive now, particularly regarding the percentage of coarse and fine particles permitted, since these particles affect melting rate and carryover. Wet batching is also quite common now to reduce batch carryover.

Glass Compositions

Table I shows changes in average glass composition over the years.[8-12] The tendency has been toward a decrease in total alkali and other fluxes and an increase in RO, primarily as straight lime. There appears to be a trend back toward the early seventeenth- and eighteenth-century glasses, which had R_2O of 3–5% and RO of up to 25%. The elimination of arsenic, nitre, fluorine, borax, and barium has been primarily due to stack-emission requirements and cost. More recent data on industry glasses seem to show a trend toward increased alkali again.

Batch cost is not the only reason for the changes that have occurred in glass compositions. Also of major importance were the chemical durability requirements imposed by the U.S. Pharmacopeia[13] and a major customer, Seagrams.[14] These specifications required higher R_2O_3, less alkali, and, in the case of distilled spirits,

36

Table I.   Glass Compositions

Oxide	1940	1950	1960	1970	1980
SiO_2	73.0	72.1	71.7	72.19	72.95
R_2O_3	1.7	1.8	2.05	1.87	1.82
CaO	8.9	8.4	11.50	9.55	10.63
MgO		2.0		1.51	0.32
BaO	0.45	0.3	0.2	0.17	0.04
Na_2O	15.0	14.4	14.50	13.96	13.65
K_2O		0.4		0.59	0.37
B_2O_3	0.5	0.2			
SO_3	0.25	0.2	0.2	0.16	0.22
F_2	0.15	0.2	0.05		

no MgO, since extraction of ppm quantities of this composition resulted in flaking of the filled container. Since few producers could produce special glasses for liquor or pharmaceuticals, the same compositions had to be used for all types of ware, large or small. Thus, the range of composition was quite wide, as each company or plant tried to fit its glasses to major product types. This range of compositions is shown in Table II.[15]

The trend toward simpler glass compositions with fewer components could result, according to some glass technologists, in glasses with poorer workability and an increased tendency toward devitrification. Operating with lower tank temperatures, to reduce fuel costs, would have the same effect.

Furnaces

Except for improvements due to the availability of better refractories and improved combustion, the original Siemans regenerative furnace concept developed about 1850 is still relatively unchanged.[16] These regenerative furnaces can be cross-fired or end-fired. Generally, small-capacity furnaces used to be end-fired up to about 100 tons/day. This is no longer true, as some end-fired furnaces are now capable of 300 tons/day.

One exception is the direct-fired Hartford unit melter, which was quite popular right after World War II and also for small production, up to about 60 tons/day.[17] These melters were extremely tricky to operate, and Tommy Coyle of Gulfport perhaps understood them better than did anyone else. At present, the unit-melter concept has been further developed by Heye in Germany to permit high pulls and complete heat recovery, and is used to produce power in the German plant.[18]

As a result of improved designs, refractories, instrumentation, and operating techniques, furnace performance measured in terms of life, tonnage (both daily

Table II.   Range of Glass Compositions

Component	1940	1950	1960	1970	1980
SiO_2	70.4–75.0	69.0–75.0	67.5–74.5	69.0–74.0	70.5–74.5
R_2O_3	0.5– 3.1	1.0– 3.0	1.5– 2.0	1.0– 5.0	1.2– 4.6
RO	5.0–10.0	7.0–13.0	9.5–12.5	9.0–13.0	9.7–12.0
BaO	0.0– 0.6	0.1– 0.3	0.1– 0.4	0.0– 0.7	0.1– 0.5
R_2O	15.0–17.0	13.0–17.0	12.75–16.75	12.5–17.0	13.2–16.0

Table III. Average Furnace Operation, 1940–1985

Year Ending Campaign	Average Furnace Life, yr	Melting Area, Cumulative ton/ft²	Average Fuel Usage, MJ/ton (Btu/ton)
1940–1945	1–1½	80– 90	10.6–12.7 $(10{-}12 \times 10^6)$
1950–1955	2–2½	120–150	9.5–11.6 $(9{-}11 \times 10^6)$
1956–1960	3½–4	336–360	8.5–9.5 $(8{-}9 \times 10^6)$
1961–1965	4½	300–350	8.5–9.5 $(8{-}9 \times 10^6)$
1966–1970	5½	360–450	6.9–7.4 $(6½{-}7 \times 10^6)$
1971–1975	6–8	540–600	6.9–7.4 $(6½{-}7 \times 10^6)$
1976–1980	8	700–750	5.8–6.4 $(5½{-}6 \times 10^6)$
1981–1985 (Projected)	9	800–900	4.8–5.8 $(4½{-}5½ \times 10^6)$

and cumulative), fuel requirements, and glass quality has changed dramatically in 40 years, as shown in Table III,[19] which gives average data from the container industry. Obviously, some of the companies had much better designers and operators than others.

A record set in 1940 showed a tank life of 1069 days or 2.9 years. This tank pulled 196.1 cumulative tons/ft² at an average of 5.3 ft²/ton (73.69 tons/day).[20] Fuel usage is not known, but was probably in the range of 10.5–11.6 MJ/ton $(10{-}11 \times 10^6$ Btu/ton). By comparison, in 1979, the record for continuous operation with *no* downtime for repairs was 4071 days or 11.15 years. This tank produced 1026 cumulative tons/ft² at 4.0 ft²/ton and 6.7 MJ/ton $(6.4 \times 10^6$ Btu/ton).[21] Fuel consumption was fairly high, as operation started in April of 1968, years before the oil embargo and when natural gas was still quite inexpensive.

Few furnaces intended for longer life and incorporating the best available technology are built today because of cost $(\$5{-}6 \times 10^6)$ and the current depressed state of industry. Instead, partial repairs are carried out to extend the tank life and improve its fuel efficiency.

Fuels have included producer gas, oil, natural gas, and partial or complete electric power, depending primarily on economics. In recent years, environmental restrictions have required the addition of various devices and techniques for limiting air pollution, as well as eliminating potentially toxic batch materials. Present batch-feeding techniques are all designed to more or less duplicate the batch patterns that resulted from hand feeding. A variety of equipment now in use includes screws, tables, and pusher and paddle chargers with both open and sealed doghouses. The use of wet batch, developed in the early 1950s, is now quite common[22]; it not only reduced carryover and batch segregation, but also increased melting rates. Numerous efforts to develop and use pelletized batch have not only been uneconomical, but have not resulted in any melting advantages or fuel

38

savings.[23] Pelletized batch may become practical if it can save fuel by permitting batch preheating.

Machines

Although the glass-container industry was already thousands of years old, it wasn't until Mike Owens invented the first truly automatic forming machine that it really began to grow into an important industry. Many more automatic machines were developed, and by the 1950s, a mixed lot were operating in the United States. These new devices included the Owens machine, various Lynch machines, the Miller, Knox I.K., Knox W.D., Roirant, and, finally, the Hartford I.S. machine.[24] During the past 40 years, virtually all of these machines have disappeared and been replaced by the I.S. machine, produced now by either Emhart or Maul. The I.S. has undergone many modifications and improvements, all of which have resulted in increased productivity. Most commonly used now are 6-, 8-, and 10-section machines with single, double, triple, and even quadruple gob capabilities. The most recent are electronically timed and program-controlled. Development in the short span of 40 years is truly fantastic: An industry which made 54 million gross of bottles in 1940 shipped 320 million gross, worth \$4.9 billion, in 1980, quite a change from \$175 million in 1940. At that time, about 126 plants, employing 67 400 people, produced 2.2 million tons of glass containers. By 1981, 20 companies employed 61 000 people and shipped 316×10^6 gross from 13×10^6 tons![25]

Annealing Lehrs

While it used to take several hours for a lehr to anneal bottles produced at rates of perhaps 20–30 tons/day, bottles produced now at a rate of 100 tons/day are annealed in 15–20 minutes. The principles of annealing, established by Adams and Williamson in 1920, still apply.[26] The improvements came because of the necessity to handle increased production rates within existing factory space. Lehrs could not be lengthened because of physical size constraints. In fact, they had to be shortened to permit the installation of single-line automatic inspection equipment. They were gradually made wider and shorter. Design changes and new control techniques permitted uniform and precise temperature distribution throughout, which resulted in shorter annealing times. Many lehrs are now electrically heated, rather than fuel-fired.

Inspection and Packing

The ever-increasing quality demanded by purchasers of glass containers forced a change in packing and inspection procedures as much, if not more, than did labor costs. As a result, 100% hand inspection and packing has been replaced over the years by ever better automatic equipment. The first devices primarily removed split and checked finishes. New equipment was continually developed until, at present, 100% automatic inspection for splits, checks, bird swings, wall thickness, stones, blisters, black spots, choke necks, down or saddle finish, and out-of-round—all with cavity identification and results printed out—exists in many plants. These devices not only discard bad bottles, but also provide information about the type, frequency, and origin of a defect, so that it can be quickly corrected. As a result, productivity and efficiency are increased.

Packing methods depend on customer demands, which vary widely; but much packing, including case packing, case palletization, bulk palletizing, and shrink or stretch wrapping of the pallets of bottles or cases, can be done automatically.

Major Problems

To attain its present size and obtain the high quality demanded by its customers, the glass-container industry has had to solve many difficult problems during the past 40 years. These problems were primarily, but not limited to, operating difficulties. Most of the answers were developed within individual companies. However, the contributions made by Preston Laboratories, later American Glass Research, particularly in the areas of strength, design, fracture analysis, and testing, were of enormous importance.

Some of the most formidable problems included

(1) Cord, both batch and surface.

(2) Excessive batch carryover and its resultant effect on checker life.

(3) Seeds and blisters, the latter particularly in amber glasses.

(4) Furnace designs for additional life, higher pull rates, and improved fuel efficiencies.

(5) Hydrodynamic, delayed, impact, and pressure breakage.

(6) Product liability, which led, finally, to investigation by the Commission on Product Safety and the development of a Voluntary Product Standard[27] for beverage containers.

Finally, competitive packages are a current problem, due to the excess weight, cost, and fragility of glass. This last problem must be overcome soon if the industry is to continue, at anywhere near its present size, into the future.

Outstanding Technical Developments

Significant developments in many specific areas permitted the glass-container industry to advance. Among these developments are composition control,[28] wet batching,[29] chemically reduced flint glasses,[30] surface treatments, deep furnace designs, automatic inspection, and computer analyses to optimize container designs. The role of the Glass Container Industry Research Corporation was also significant. Among its accomplishments were the investigation of pelletized batch, the development of the first electronically timed machine, high-speed photography of the forming and filling processes, and submerged burners for melting and forehearth coloring.[31] Chemical strengthening, although a significant accomplishment, was ultimately uneconomical, and had no future as originally conceived.[32] Plastic encapsulation had a negative effect on the glass-container industry, as it paved the way for the introduction of large plastic beverage containers. Also on this list of great accomplishments, perhaps ahead of their time, was the GCMI 2000 bottle/minute filler.[33] In addition, the GCP package developed by O-I was perhaps the forerunner of the current single-service container, which permitted the glass industry to maintain its share of the soft-drink market, in competition with cans and plastic.[34] In 1981, for example, 61.9% of all containers shipped were beverage bottles for soft drinks, beer, wine, and liquor; soft drinks were 18–19% of that total.[35]

The control of durability after processing by using a specific fluorine treatment has also had a significant effect, as chemical durability no longer must be controlled by the original glass composition.[36] Workability problems, primarily due to phase separation or microcrystalline residues,[37] have also decreased over the years. Greatly improved forming machines today provide high productivity, computer control, and mold-cooling systems based on first principles.

Undoubtedly, many more outstanding developments have not been included in this list. Certainly, one of these developments would be the ability of the glass industry, through knowledge of design, strength, and product requirements, to have continually reduced package weight, while still retaining adequate strength for the proposed use. The greatest research effort, however, will be to carry this particular process much farther, so that the glass container can compete with a growing host of packaging materials.[38]

A discussion of major problems and outstanding developments in the glass-container industry over the past 40 years would not be complete without some mention of the significant impact that the Glass Problems Conferences have had on the solutions to practical problems. These conferences were organized by Fay Tooley and Henry Blau in the late 1930s and have been held in alternate years at the University of Illinois and the Ohio State University. There has always been a strong emphasis on the solution of practical problems at these conferences, and they were typically attended by the individuals responsible for production, engineering, furnace operation, and control laboratories, as well as representatives of various refractory and material suppliers. Many feel that the relaxed atmosphere in the hospitality suites of these suppliers provided the basis for a useful information exchange that could have been accomplished in no other way. However, the emphasis on the solution of practical problems at these conferences was in itself a symptom of a basic problem for the glass-container industry. This basic problem involves the relative lack of communication and interaction between those people involved in the day-to-day problem solving and those in the academic community who do fundamental glass research. For many years, a lack of communication and coordination between these groups resulted in poor transfer of developing technology to the various processes involved in the industry. Current developments, it is hoped, will address this problem: Certainly, this session at an American Ceramic Society meeting is a case in point.

Conclusion

The growth of the glass-container industry since 1940 is truly phenomenal and has resulted from major expenditures of money and individual effort toward the solution of problems and the evolution of technology. While the glass-container industry presently faces great competition from other packaging materials, there is also a growing realization in the industry of the kind of research and development necessary to maintain or increase its market share against these competitive forces. A number of basic research projects for improving the practical strength of glass by several orders of magnitude now exist in various universities throughout the world. It is success in this type of effort that will provide the best opportunity for glass to continue as the premier packaging material.

References
[1]J. Small, "Walking Watchfully through a Buyers Market," Glass Tech., 24 [5] 224–30 (1983).
[2]L. G. Ghering, "Refined Method of Control of Cordiness and Workability of Glass during Production," J. Am. Ceram. Soc., 27 [12] 373–87 (1944).
[3]M. A. Knight, "Cords in Glass," Glass Ind., 37 [9] 491, [10] 553, [11] 613, [12] 668 (1956).
[4]"Current Statistical Position of Glass," Glass Ind., 22 [2] 70 (1941).
[5]"The Glass Industry Goes on Trial," Glass Ind., 21 [1] 11–12 (1940).
[6]M. J. Owens, "Glass Shaping Machine," U.S. Pat. No. 766 768, August 2, 1904.
[7]W. Trier and W. Giegerich; pp. 260–313 in Glass Machines. Springer-Verlag, New York 1969.
[8]D. E. Sharp, "Recent Trends in Glass Composition," Glass Ind., 21 [4] 160 (1940).
[9]A. L. Bracken, Jr., "A Review of Container Glass Compositions, 1940–1950," Glass Ind., 32 [9] 449–53 (1951).
[10]R. E. Loesell and W. R. Lester, "Container Glass Compositions, 1932 to 1960," Glass Ind., 42 [11] 623–29 (1961).
[11]Leo E. Stadler and A. W. LaDue, "Review Shows Status of Glass Compositions," Glass Ind., 53 [12] 14–16 (1972).

41

[12]E. McKinley; private communication.
[13]"Chemical Resistance in Glass Containers"; pp. 924–25 in U.S. Pharmacopeia XVIII. Mack Printing Co., Easton, PA, 1970.
[14]A. Herman, "Factors Influencing Autoclave Chemical Durability Tests of Glass Containers," J. Am. Ceram. Soc., 24 [10] 323–27 (1941).
[15]A. Herman, "Factors Influencing Autoclave Chemical Durability Tests of Glass Containers," J. Am. Ceram. Soc., 24 [10] 323–27 (1941).
[16]Karl Wilhelm Siemans, Encyclopedia Americana, Vol. 24. 1941.
[17]G. E. Howard, "Method of Making Glass," U.S. Pat. No. 1 999 762, April 30, 1933.
[18]"New Heye Melter Commissioned at Obernkirchen Glassworks"; pp. 52–53 in Glass International, 1983.
[19]Brockway, Inc. (NY); private communication.
[20]Corhart Refractories Co., Glass Ind., 22 [5] 187 (1941).
[21]Brockway, Inc. (NY); private communication.
[22]J. P. Poole, "Glass Batch and Method for Preparing Same," U.S. Pat. No. 2 813 036, November 12, 1957.
[23]R. Miller and H. Moore, "Compacted Batch—Will It Make a Difference?" Glass Ind., 60 [6] 20–23 (1979).
[24]W. Trier and W. Giegerich, Glass Ind., 60 [6] (1979).
[25]Glass Packaging Institute Annual Report, 1982.
[26]L. H. Adams and E. D. Williamson, J. Franklin Inst., 190, 597–631, 835–70 (1920).
[27]Voluntary Product Standard PS73-77, "Carbonated Soft Drink Bottles," U.S. Dept. of Commerce/Nat. Beer Stds., September 15, 1977.
[28]J. P. Poole, "Practical Glass Composition Control," Glass Technol., 25 [2] 76–82 (1984).
[29]J. P. Poole, "Practical Glass Composition Control," Glass Technol., 25 [2] 76–82 (1984).
[30]J. P. Poole, Fundamentals of Fining. International Commission on Glass, Toronto, Canada, 1969, pp. 169–76.
[31]Glass Container Industry Research Corporation (confidential reports).
[32]J. P. Poole, H. C. Synder, and M. A. Boschini, "A Method of Strengthening Glass and Increasing the Scratch Resistance of the Surface Thereof," U.S. Pat. No. 3 743 491, July 3, 1973.
[33]Glass Packaging Institute Annual Report, 1971.
[34]H. E. Simpson, "The Glass Industry—1969," Glass Ind., 51 [4] 160–61 (1970).
[35]Glass Packaging Institute Annual Report, 1982.
[36]J. P. Poole, H. C. Snyder, and R. J. Ryder, "Corrosion Retarding Fluorine Treatment of Glass Surfaces," U.S. Pat. No. 3 314 772, April 18, 1967.
[37]J. P. Poole, "Glass Workability," Glass Ind., 48 [3] 129–36 (1967).
[38]R. S. Coakley, "Glass Can Achieve Its Strongest Position Yet," Am. Glass Rev., 105 [7] 12–13 (1984).

Float Glass

CHARLES K. EDGE

PPG Industries, Inc.
Glass Research and Development
P.O. Box 11472
Pittsburgh, PA   15238

The production of glass by float-forming processes has been the most significant development in the manufacture of flat glass to occur in the past fifty years. A brief description of the two commercial float-glass processes currently being practiced is given, and the old patent literature is used to construct a "prehistory" of float-forming. Two physical phenomena (the concept of an equilibrium thickness and the decay of waves on a viscous surface) which have direct impact on the processes are reviewed, and some misinterpretations of these effects which have occurred in the past are discussed.

The development and successful commercialization of float-glass processes for the manufacture of flat glass have totally transformed this industry over the past quarter of a century. In the United States, plate-glass production is an extinct technology, and sheet-glass production is nearly so. Worldwide, the supplanting of these technologies by float processes is a continuing and apparently irreversible phenomenon.

For background, a brief description of a float-forming process is appropriate at this point. The glass tank which melts and conditions the molten glass prior to forming is (typically) 30 feet wide, at least 150 feet long, and usually contains more than 1200 tons of glass. The most common source of fuel is natural gas (with fuel oil also used on occasion), and the product is of high metal quality, with little ream and less than one point defect in every 1860 m² (200 ft²) of the final ribbon.

The glass is cooled in the front of the tank to 1093°C (2000°F) or less and conveyed to the forming chamber, which contains a pool of molten tin and is the heart of a float-forming process. On this molten substrate, the ribbon flows to an equilibrium thickness and is further reduced in thickness by the action of attenuating forces from the lehr (and, in most instances, sizing machines, which grip the ribbon on its edges). This combination of forces produces the desired final thickness: The ribbon then cools until it is dimensionally stable, whereupon it is conveyed to an annealing lehr.

There are currently two float-forming processes in commercial use, one developed by Pilkington Brothers (PB) and one by PPG. There are major differences between the two processes, not the least of which are the means by which the glass is delivered to the forming chamber and the procedures and philosophies used in sizing the ribbon. Figure 1 is a schematic of the PB process; Fig. 2 depicts the PPG process.

Despite the commercial success of float, the technical literature contains relatively few contributions to the science and technology of float-glass forming,

43

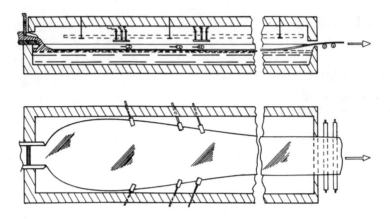

Fig. 1. The PB float process.

due, in large part, to understandable proprietary considerations. The standard
reference for the earlier work on the PB process is an article by Pilkington[1]; it has
been supplemented by more recent reviews by Edge[2] and Hynd.[3] In addition, an
article by McCauley[4] compares the PPG and PB processes.

Even though the amount of open literature on float is relatively sparse, the
many patents on float-forming are a good source of information about the forming
of glass on a molten-metal substrate, although the usual caveats concerning their
interpretation apply.

A situation of this type produces some unusual variations on the general theme
of technological development, even within a given corporation. Significant sci-

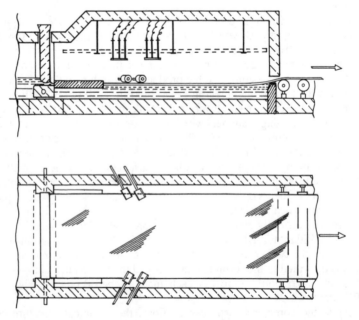

Fig. 2. The PPG float process.

entific and engineering successes can, and usually do, parallel areas which remain relatively backward. Certain practices and beliefs are based on folklore or tradition, and, in general, development is rather uneven when perceived from an interdisciplinary point of view.

Using the patent literature, a "prehistory" of float-forming can be developed; by using the available technical literature, some bits of folklore can be analyzed more critically. This article will look at both points in more detail.

Prehistory

The first patent to disclose the use of a molten-metal support in the manufacture of glass was granted to Bessemer in 1848.[5] In this document, Bessemer anticipated certain technological aspects of modern float manufacture, such as:

(1) The use of molten tin as a support metal.

(2) The possibility of dross formation on the exposed metal surface.

(3) Such practices as using a protective reducing atmosphere over the molten metal, operating the chamber at a positive pressure, and a method for removing dross.

Nevertheless, the Bessemer patent does not really address the forming process per se. To quote from the patent, "... my invention consists of preserving the flatness of sheets and plates of glass during the annealing process by using a fluid instead of a solid substance to support them upon."[5] In other words, Bessemer is talking about the annealing, not the forming, of glass on molten metal.

The first "float" patent was probably that granted to Luigi Lombardi in 1900,[6] and it did not involve glass at all: Lombardi described a process for manufacturing flat plates of a dielectric material (for use in capacitors) by pouring a viscous liquid such as wax or paraffin onto another liquid such as mercury and allowing the upper fluid to congeal and form a flat plate.

The first patent describing the formation of glass on a molten-metal substrate was issued to William Heal in 1902.[7] Heal described a process where glass was continuously delivered onto a molten metal (tin was mentioned specifically). The production of various thicknesses by applying a tractive force to the ribbon was revealed as well. In the Heal patent, metal oxidation was avoided by having the glass completely cover the support liquid.

In 1905, H. K. Hitchcock[8] discussed the extrusion of molten glass onto a liquid metal compartmentalized for temperature control: The production of various thicknesses was accomplished by what was essentially an embryonic version of the direct-stretch process being practiced in float baths today.

Other patents were issued to Heal[9] on the melting of powdered cullet directly on a molten tin support to form a glass plate and to Hitchcock[10] for a process in which glass at a high temperature is poured onto a liquid, allowed to spread in a lateral direction, and then rolled to the desired thickness.

There is a gap of 23 years between the two Heal patents and one of 20 years between those of Hitchcock. The clear implication is that the forming of glass on a molten-metal substrate was not just a passing fancy. Accordingly, two questions come to mind:

(1) Was any experimental work on float-forming performed during this period?

(2) Why did the concept not develop further in the first half of this century?

It is probably impossible to answer the first question definitively. However, an experiment which *may* have been the first was done at PPG's Creighton plant in the late 1920s. A small basin to contain the molten metal was built onto the side of an operating plate-glass tank, which served as the source of glass. A sheet was apparently formed, but the experiment was terminated when the basin cracked, and the metal was lost. The experiment was not repeated. An interesting feature of this work was that the molten-metal support was not tin, but antimony.

The reasons for the lack of commercial development at this time seem a bit more obvious. Flat-glass manufacture in the 1920s saw two major breakthroughs, with glassmelting transformed from a batch process to a continuous one, and the introduction of the various sheet-forming processes such as the Colburn, Fourcault, and Pittsburgh techniques. Given the obvious fact that all glass companies, now as well as then, have only finite quantities of available capital and personnel for R&D, the active pursuit of float-forming, which needed the continuous melter, but competed directly with plate- and sheet-glass forming, would understandably have been given a low research priority, even assuming that support for such development existed.

The Breakthrough

The announcement of the successful demonstration of their float-glass forming process by Pilkington Brothers some 25 years ago culminated almost a decade of intensive developmental work. The history of this effort has been admirably recounted by Pilkington,[1] and it would serve little purpose to repeat it here, but a few comments are in order: First, one can only admire the tenacity shown by Pilkington's technical staff throughout the long years of work; in addition, the financial support provided by management shows no small amount of patience and economic courage. Second, the change in the process itself can be traced from the original concept of a finishing step for plate glass, to a direct-pour process for replacing plate, to a process for producing glass of all commercial thicknesses (although this latter phase did not really take place until the early 1970s). Finally, while the scientific achievement associated with the development of float was considerable, particularly in the field of float-bath chemistry, it is at least arguable that the success of float glass is due more to engineering accomplishments than to scientific results.

Surface Tension and Wave Decay

As stated previously, the original motivation behind Pilkington's development[1] was to provide a process for fire-polishing a ribbon of rough-rolled plate glass, thus eliminating the grinding and polishing steps. The first tin bath was apparently built onto the front of a plate-glass furnace. When a ribbon $\approx$6-mm thick was introduced onto the bath the process worked, and a fire-polished ribbon was produced. But then a curious thing happened: A 3-mm-thick plate ribbon was fed to the bath, the ribbon was fire-polished, and it emerged half as wide and 6-mm thick! The process produced a ribbon whose final thickness was independent of its initial thickness. What, of course, had been demonstrated was a high-temperature example of the surface tension of two immiscible liquids producing a lens (in this case the float ribbon) whose equilibrium thickness was a function of the physical properties of the two fluids (and the atmosphere) alone.

The classic reference on this particular phenomenon is a paper by Irving Langmuir[11] which was published in 1933. Figure 3 shows this equilibrium thickness concept; the formula is from Langmuir's paper, and its success in predicting the proper value for the equilibrium thickness of glass floated on tin vindicates the theory.

There is, however, a problem. A paper by Pujado and Scriven[12] appearing in 1972, among other things "corrects serious errors in Langmuir's 1933 analysis,"

46

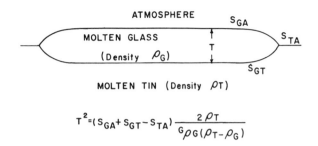

$$T^2 = (S_{GA} + S_{GT} - S_{TA}) \frac{2\rho_T}{G\rho_G(\rho_T - \rho_G)}$$

Fig. 3. Vertical section of a pool of molten glass floating on molten tin.

and derives the appropriate formulas for various lens-forming situations. (The float ribbon, in their terminology, is a "translationally symmetric, semi-infinite, sessile lenticular configuration.")

As far as can be determined, this paper has never been cited in the glass and ceramic literature. In fact, a citation analysis of both papers is of some interest: Table I shows such an analysis for the years 1974–1984.

Several comments are in order. It should be remembered that the Langmuir paper is predominantly experimental, dealing with the formation of lenses of high-hydrocarbon oils on a water surface. It is a tribute to its enduring value that it is cited so frequently decades after its publication. Pujado and Scriven[12] made a contribution that is mostly theoretical, but which has also become established in the literature.

What is the source of the discrepancy between these two papers? It is that Langmuir, in deriving his equilibrium thickness equation, made some unwarranted assumptions, which resulted in his obtaining the right formula for the wrong configuration!

This confusion about surface tension has direct impact on such areas as the analysis of the float-forming process, where the surface-tension force has been included both isotropically and anisotropically. The latter approach is probably correct and is exemplified, or at least implied, in a recent paper by Popov.[13]

Another subject area important to float-glass manufacture is that of the optical distortion present in the final product. This distortion is primarily due to small lateral thickness variations in the float-glass ribbon that form positive and negative lenses and are characterized by a specific wavelength of ≈5 cm. Thus, it is a natural step to analyze the decay of these waves (or the surface flattening of the ribbon), in an effort to determine which conditions would further reduce the distortion levels present in the final product.

Table I. Citation Analysis (1974–1984)

	Langmuir*		Pujado and Scriven†
Number of citations	46		11
Number of citations in common		2	
Number of citations (glass and ceramic literature)	2		0

*Ref. 11.
†Ref. 12.

47

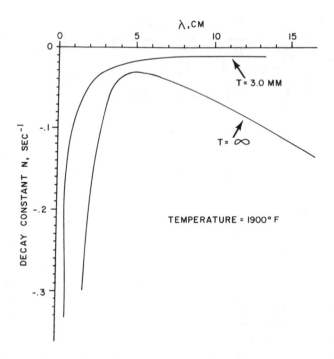

Fig. 4.   Wave decay at a glass/atmosphere interface.

The literature is, again, quite sparse on this subject, although the first treatment of the problem appeared a century ago in a book by Basset.[14] His analysis shows that the decay of waves on the surface of a viscous liquid proceeds by two mechanisms: The long wavelengths decay due to gravitational forces, whereas very short wavelengths decay because of surface-tension effects. The combination of the two produces a critical wavelength, which decays at the slowest rate. The hydrodynamic literature of the century since Basset's analysis contains an enormous amount of work on wave decay, but virtually none of it is concerned with viscous effects.

Basset's theory was applied to glass in 1969, in an excellent paper by Woo,[15] where experimental data appeared to justify the analysis, which predicted the correct critical wavelength for soda-lime glass.

This conclusion is, however, misleading because both Woo and Basset assumed that the viscous fluid in question was of infinite depth. This enabled a straightforward solution of the appropriate secular equation. They both also formulated the corresponding problem for a sheet of finite thickness (though in slightly different ways) and derived, but did not solve, the equations for this case. Further analysis of this problem[16] yields some quite different conclusions: Wave decay at a given temperature is much slower than predicted by Basset's theory, because the gravitational mechanism of decay is inoperative, and surface tension is the only wave-decay mode available; as a result, there is no critical wavelength. These effects are important, and in fact dominant, for a glass thickness of 3 mm or less. Figure 4 compares these two cases. More details relevant to these points will be published later.

Another curious point concerning the wave decay phenomenon is that waves are decaying on two glass surfaces, and it might be asked why the critical wavelength for decay is the same at a glass/tin interface as at the glass/atmosphere interface. The answer, based on analyses by Rayleigh,[17] Chandrasekhar,[18] and Hide,[19] is that, for glass, equality of critical wavelengths is a numerical accident, due to the equality of the ratio of the coefficient of surface tension to the density difference at each interface.

Conclusion

As stated earlier, a history of the development of float-forming processes, especially during the past two decades, can be found elsewhere,[2,3,20] and, in general, more papers are now appearing in the open literature. It is clear that float-glass technology is still in a stage of continual development and that many scientific and technological challenges remain (for example, in the areas of float-bath chemistry and the analysis of the forming process). These remaining challenges should ensure a fertile climate for research and development in float glass well into the future.

Acknowledgments

The author would like to thank H. R. Foster, F. J. Matjasko, and K. A. Sobotka for their help in preparing this paper.

References

[1]L. A. B. Pilkington, "The Float Glass Process," *Proc. R. Soc. (London), Sect. A,* **314**, 1–25 (1969).
[2]C. K. Edge, "Flat Glass Manufacturing Processes (Update)"; pp. 714-1–21 in Handbook of Glass Manufacture, Vol. II. Edited by F. V. Tooley. Ashlee Publishing Co., Inc., New York, 1984.
[3]W. C. Hynd, "Flat Glass Manufacturing Processes"; pp. 83–100 in Glass: Science and Technology, Vol. 2. Edited by D. R. Uhlmann and N. J. Kreidl. Academic Press, New York, 1984.
[4]R. A. McCauley, "Float Glass: Pilkington vs PPG," *Glass Ind.,* **61** [4] 18–22 (1980).
[5]H. Bessemer, "Manufacture of Glass," Br. Pat. No. 12 101, September 22, 1848.
[6]L. Lombardi, "Process for Manufacturing Thin Homogeneous Plates," U. S. Pat. No. 661 250, November 6, 1900.
[7]W. Heal, "Manufacture of Window and Plate Glass," U. S. Pat. No. 710 357, September 30, 1902.
[8]H. K. Hitchcock, "Apparatus for Manufacturing Glass Sheets or Plates," U. S. Pat. No. 789 911, May 16, 1905.
[9]W. Heal, "Method of Making Glass Plates," U. S. Pat. No. 1 553 773, September 15, 1925.
[10]H. K. Hitchcock, "Process and Apparatus for Making Sheet Glass," U. S. Pat. No. 1 564 240, December 8, 1925.
[11]I. Langmuir, "Oil Lenses on Water and the Nature of Monomolecular Expanded Films," *J. Chem. Phys.,* **1**, 756–76 (1933).
[12]P. R. Pujado and L. E. Scriven, "Sessile Lenticular Configurations: Translationally and Rotationally Symmetric Lenses," *J. Colloid Interface Sci.,* **40**, 82–98 (1972).
[13]V. V. Popov, "Spread of Molten Glass on the Surface of Liquid Tin," *Fiz. Khim. Stekla,* **7**, 717–22 (1981).
[14]A. B. Basset; pp. 309–13 in Hydrodynamics, Vol. II. Deighton, Bell and Company, 1888.
[15]T. C. Woo, "Wave Decay on Glass Surface at High Temperatures," *J. Appl. Phys.,* **40**, 3140–43 (1969).
[16]C. K. Edge; unpublished analysis.
[17]Lord Rayleigh, *Proc. Math. Soc. (London),* **14**, 170–77 (1883).
[18]S. Chandrasekhar, "The Character of the Equilibrium of an Incompressible Heavy Viscous Fluid of Variable Density," *Proc. Camb. Philos. Soc.,* **51**, 162–78 (1954).
[19]R. Hide, "The Character of the Equilibrium of an Incompressible Heavy Viscous Fluid of Variable Density: An Approximate Theory," *Proc. Camb. Philos. Soc.,* **51**, 179–201 (1954).
[20]R. C. Perry, "The Float Process for Manufacturing Flat Glass"; pp. 254–59 in the XIII International Congress on Glass, Vol. 1. Deutschen Glastechnischen Gessellschaft, 1983.

Glass Fibers

P. F. AUBOURG AND W. W. WOLF

Owens Corning Fiberglas Corp.
Granville, OH   43023

After reviewing typical applications of glass fibers and the technologies used to produce them, requirements imposed on the glass by-product and fabrication constraints are discussed. The composition and properties of various types of glasses developed in the past are discussed as well as recent compositional trends.

In the space of about fifty years, glass fibers have grown from a novelty product with little commercial application into a material whose production is now measured in millions of tons annually. Glass compositions used in manufacturing fibers for wool insulation and for reinforcement have evolved over the years. Optical fibers will not be covered here, but an overview of the glass compositions used in the fiberglass industry and their development will be given. Glass composition is determined by such product requirements as strength, water durability, refractive index, dielectric constant, etc.; by manufacturing requirements; and by economic requirements, i.e., cost. For this reason, a brief overview of product applications and manufacturing techniques follows.

Product Applications

The range of applications for fiberglass has grown from a single product in the 1930s — a fiberglass mat for air filters — to tens of thousands of applications at present. Table I gives a cursory list of fiberglass applications, as well as some of the glass requirements for these applications. Applications have been regrouped into three categories: (1) those using discontinuous fibers packed as batts or boards; (2) those using fabrics and mats; and (3) those for reinforcement, using continuous or chopped fibers.

Thermal and acoustical insulation are the main applications of discontinuous glass fibers. These products are used extensively in buildings, transportation equipment, and appliances. The glass must have adequate mechanical strength and chemical durability. Resistance to water attack is important, due to the size of the fibers. Typically, the average diameter of wool-glass insulation fibers is 5 to 6 μm. An early glass-composition research program conducted at Owens Corning Fiberglas Corp. (OCF) tried to solve a problem encountered with wool insulation used in railroad refrigerator cars: The combination of condensation and vibration degraded the wool insulation. Development of glasses with improved durability solved this problem.

The second category, fabrics and mats, covers all applications where fiberglass protected by an organic or metallic coating is used by itself, i.e., not in a matrix. For example, glass mats are used in batteries to prevent shorting of the electrodes during discharge. This glass must, therefore, have exceptional resistance to mineral acids.

Table I. Fiberglass Applications

	Glass Properties
Discontinuous fibers	
Insulation — thermal, acoustical	T, fib. dia.
Filters	T, chem.
Fabrics and mats	
Drapes, fabrics	fib. dia., mech.
Clothing	T, fib. dia., mech.
Cable and wire insulation, sleeves	elec.
Yarn and threads, screening	mech.
Battery-retainer mat	chem.
Reinforcement — continuous and chopped fibers	mech.
Plastics and polymers —	
boats, automotive, aircraft, consumer products	(chem.)
storage tanks	chem.
appliance housing	elec.
printed circuit boards, electronic	elec.
transparent panels	optic.
Teflon* — fabric structures	fib. dia.
rubber — tire cord, tires	
asphalt — shingles, road bond	
plaster, gypsum	
cement, mortar	chem.

Note: T = high temperature resistance; chem. = chemical durability; mech. = mechanical properties (modulus, strength, etc.); elec. = electrical properties (resistivity, dielectric, etc.); optic. = optical properties (color, refractive index, dispersion); and fib. dia. = controlled fiber diameter and/or fine fibers.
*E. I. du Pont de Nemours & Co., Wilmington, DE.

The third category, continuous or chopped fibers used as reinforcement, covers very diverse markets. These applications take advantage of the high specific strength of glass combined with (depending on the application) chemical inertness, low electrical conductivity, low dielectric constant, or transparency. For all of these applications, the glass is coated with an organic size, or finish, that ensures good bonding between the glass and the material to be reinforced.

The glass properties needed for specific applications are listed in Table II, as are the glasses that have been developed over the years to meet these needs. It should be pointed out at this time that, due to the very high cooling rate of the glass in the fiberizing process (of the order of 10^5 °C/s), fiberglass properties differ slightly from those of bulk annealed glass: Density, refractive index, and Young's modulus are lower for the fibers.

At an early stage in the development of fiberglass, a glass with low electrical conductivity (bulk and surface conductivity) was required for electrical applications. The standard soda-lime-silica glass composition for sheet glass, or A glass, could not meet this requirement due to its high alkali level. This led to the development of E, or electrical, glass. To meet conductivity requirements, the alkali level must be less than 2 wt% and is, in practice, less than 1 wt% for many applications. This glass has, at the same time, excellent water durability, good mechanical properties, and a refractive index matching available resins; for these reasons, it has become the standard glass for most continuous-fiber applications.

Table II. Glass Property Requirements and Glasses Developed to Meet These Needs

		Glass Types
Electrical properties	Conductivity	E
	Dielectric constant	
Mechanical properties	Strength	S
	Modulus	(M)
	(Density)	
Chemical durability	Water	E, wool
	Acid	C
	Acid/leachable glass	
	Base/alkali	AR
Optical properties	Refractive index	(W1), W2
	Color	
Special properties	Radiation absorption	(L)

Note: Glasses enclosed in parentheses are not commercially available in large quantities.

The letters used in Table II do not represent one glass composition, but rather a range of compositions.

The need for higher-strength glasses led to the development of S (for strength, or strong) glass. The critical property in many applications is not strength itself, but strength for a given weight, or specific strength. For this reason, density is listed with the mechanical properties in this table. For many structural applications, the limiting design factor is stiffness rather than strength. This led to the development of a high-modulus glass, or M glass.

Durability is a critical property for many applications, and E glass has outstanding water durability. As discussed before, the wool-glass composition was modified to improve water durability. Manville Industries uses a highly chemical-resistant glass for their very fine fiber insulation and filtration product. Since E glass has poor acid resistance, a chemically resistant, or C, glass was developed to meet these needs. The poor acid durability of E glass was turned into an advantage to produce high-silica (>96%) fiber by a process similar to the one used for Vycor.* The E glass can be acid-leached to a silica network without prior heat treatment. Glasses with improved leachability have been developed, but are not commercially produced due to the limited size of the market. To reinforce cement, alkali-resistant, or AR, glasses were developed. Commercially available AR glass has outstanding resistance to corrosion by acids as well as bases.

Color is critical for clear panel applications. Low-color, E-glass modifications were developed for this market and are referred to as W1 and W2 glasses.

A high-lead glass fiber, or L glass, was developed to absorb radiation.

As can be seen from this partial list, glasses have been developed to respond to many needs. The bulk of the production of glass fibers is, however, limited to a few of these glasses: wool compositions for insulation, E-glass compositions for reinforcement. Other glasses such as C, S, and AR are sold for specific, narrower markets. A controversy exists on the merits of using sheet-glass composition, or A glass, instead of E glass for nonelectrical applications.[1] This glass was used in the 1960s in Europe, mostly in the United Kingdom. As will be seen later, this glass does not have comparable water durability and is more sensitive to stress

---

*Corning Glass Works, Corning, NY.

Table III.   Manufacturing Constraints on Glass Composition

Step in Manufacturing Process	Constraints
Batching	Raw materials availability
Melting	Refractory life/refractory corrosion Melting energy
Fiberizing	Glass-forming viscosity compatible with available metals Glass-crystallization tendency Glass quality
Conversion into usable product	
Overall process	Economically acceptable cost

corrosion. For this reason, little effort has been made to commercialize A glass in the United States.

Manufacturing

The steps involved in manufacturing glass fibers, as well as the constraints imposed on glass composition by each step, are listed in Table III. This discussion will be limited to constraints imposed by the fiberizing process, since the first two steps, batching and melting, are common to the whole glass industry. As indicated before, there are two types of glass fibers, discontinuous and continuous, and it is necessary to consider them separately in describing manufacturing processes.

Discontinuous Processes

These discontinuous processes are used not only to produce insulation and filter products, but also to make staple fiber or sliver. This last application is now limited. Many processes have been used, and a few are described here.

(1) Steam and air-blown process — this process is an improved version of the old mineral wool process: Small streams of molten glass are fiberized by high-pressure steam or air jets that break the glass stream into discontinuous fibers. This process gives fairly coarse fibers with a high shot (unfiberized beads of glass) content. The resulting product is, therefore, of fairly poor quality, but glass requirements are minimal. Glasses with very high viscosity and liquidus can be fiberized by this process.

(2) Mechanical attenuation — in this process, the glass stream is attenuated by falling on a series (generally three) of rapidly rotating wheels. Again, this process generates shots, and glass requirements are minimal.

(3) Flame attenuation — this fairly old process is still in use. Primary fibers, approximately 1 mm in diameter, are drawn through bushings in the bottom of the glass tank. These filaments are then aligned in an exact, uniformly juxtaposed array and exposed to a jet flame from an internal combustion burner. The flame breaks the filaments into long fibers with narrow diameter distribution that pass into the binder spray. Very fine fibers can be produced. The glass requirements for this process are similar to those for the continuous process, but are not as stringent due to the larger diameter of the primary fibers. A recent modification is the toration process.

54

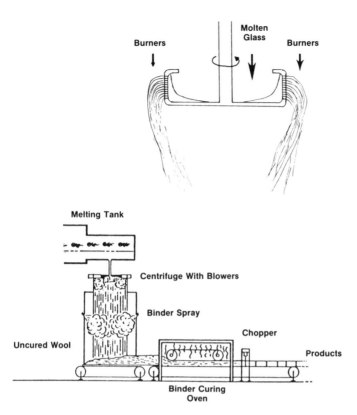

Fig. 1. Rotary process for discontinuous fibers.

(4) Rotary process — this procedure is now the most commonly used. In this process (see Fig. 1), the molten glass from the melting tank drops into a rotating cylindrical unit, or spinner, that has numerous holes in its sidewalls. Centrifugal forces extrude the molten glass, and it streams laterally from the holes into the path of a high-velocity gas stream (air, steam, or combustion gases) that attenuates and breaks it into fine, discontinuous fibers. This is a high-throughput process that has the advantage of producing no shot, but it places more demands on the glass. The fiberizing temperature and, hence, viscosity must be low enough to meet the mechanical limitations of spinners. The liquidus must also be significantly below the fiberizing temperature to prevent crystal growth in the spinners and plugging of the holes.

Continuous Processes

Two basic continuous processes are used: marble melt and direct melt. The direct-melt process is more energy efficient, since reheating is not required. However, marble melt is more adaptable, since the marbles can be shipped from a central point to many different fiber-producing plants.

Commercial glass fibers are made by extruding molten glass through an orifice, usually 0.8 to 3 mm in diameter, and then rapidly pulling this glass to draw it to a fine diameter, from 3 to 20 μm. Mechanical winders to do this drawing operate at speeds up to 220 km/h (see Fig. 2).

55

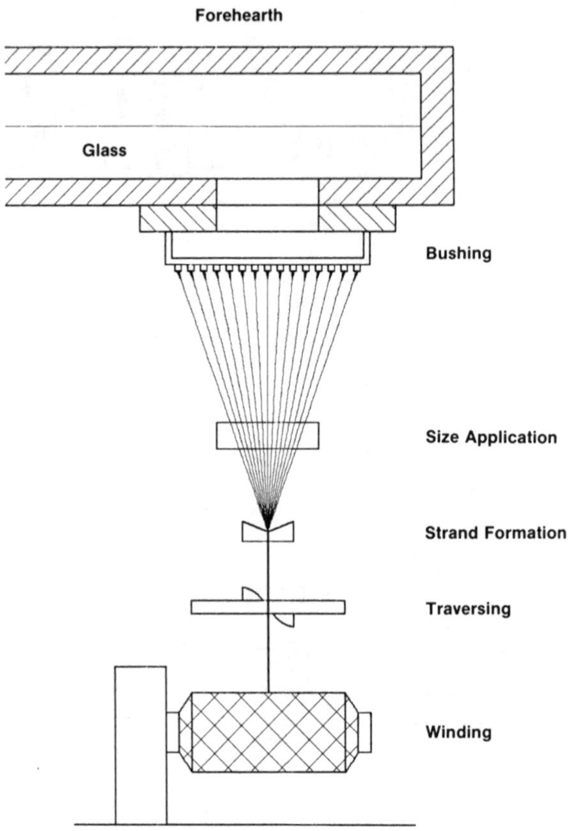

Fig. 2. Continuous process.

The orifices through which the molten glass passes for fiberizing are in the base of a platinum-alloy "bushing," which is heated electrically. Most bushings carry 204 orifices or some multiple of this number. The individual filaments are combined into one strand, which is the basic building block for glass-fiber products. Glass filaments are highly abrasive to one another; to minimize degradation of filament strength due to abrasion, size coatings are applied before the strand is gathered.

The glass properties needed for ensuring efficient fiberization are

(1) High glass quality

Inclusions — the diameter of continuous glass fibers ranges from 3 to 20 μm. In such fine filaments, solid inclusions of even submicrometer dimensions act as stress concentrators. Such inclusions could be specks of refractory dislodged from a sidewall due to a change in level, a small piece of devitrified glass arising from too low a temperature in a forehearth, or metal dust rubbed off the feeding tubes into a marble bushing; they all cause breakage of a single filament in the first instance, the tail of which then breaks others and leads to an interruption of the fiber-drawing process.

Homogeneity — inhomogeneity, such as could arise from inadequately mixed raw materials, or the inadequate dissolution and dispersal of striae of alumina or

Table IV. Wool-Glass Compositions, Wt%

	A Glass	T Glass (Bowes, 1939 (Ref. 2))		(Welsch, 1955 (Ref. 7))	
SiO_2	72	60–65	63	50–65	58.6
Al_2O_3	2 ⎫	2–6	5	0–8	3.2
Fe_2O_3	⎬			0–12	
CaO	5.5 ⎫	15–20	14	3–14	8.0
MgO	3.5 ⎬		3	0–10	4.2
BaO				0–8	
B_2O_3		2–7	5	5–15	10.1
$Na_2O + K_2O$	16	8–12	10	10–20	15.1
ZnO				0–2	
MnO				0–12	
TiO_2				0–8	
ZrO_2				0–8	
F_2					
T_3 (°C)*	≈1200		1050		1015
Liquidus (°C)	≈ 850		1160	870–980	925

*T_3 = Temperature in °C at which the glass viscosity is 10^3 P.

silica, causes local sudden changes in viscosity which can increase or decrease the supply of glass through nozzles outside the range of successful fiber formation, thus also leading to filament breakage during fiber manufacture.

(2) Low liquidus

This property is a corollary of the need for high glass quality. Temperature fluctuations inherent to the process must not lead to glass crystallization and the formation of small crystals that would induce breaks.

(3) Low viscosity, to limit deformation of the precious metal bushings

Typically, the glass is fiberized at viscosities of 500 to 1500 poises. The need for low forming temperatures is best illustrated by an example: S glass is fiberized at temperatures several hundred degrees centigrade higher than is E glass and, as a result, the lifetime of a bushing used to fiberize S glass is less than 20% that of an E-glass bushing.

Glass Compositions

Discontinuous Fibers

A typical soda-lime-silicate composition is given in Table IV (Glass A). These glasses were rapidly replaced by T, or thermal, glasses with improved water and acid durability and lower viscosity. The glass has a comparatively high liquidus, which was acceptable for the Owens steam-blowing process. The composition range of this glass is covered in the Bowes patent.[2] Patented range and recommended composition are given in Table IV. The viscosity was reduced by adding CaO and B_2O_3 and reducing SiO_2; durability was increased by adding B_2O_3 and Al_2O_3 and reducing Na_2O; finally liquidus and crystallization tendencies were reduced by adding Al_2O_3 and B_2O_3 and keeping the ratio of MgO/CaO to around 0.25.

Over the years this glass was modified to reduce batch cost and viscosity. Boron was, as now, the most expensive raw material. The Dingledy patent[3] describes a composition with lower viscosity obtained by increasing alkaline earth content at the expense of SiO_2. Durability was maintained at an acceptable level by adding TiO_2. A low-cost modification of this glass was developed by Tiede.[4] To reduce cost, TiO_2 and B_2O_3 levels were kept very low; as a result, glass durability was not as good. Product properties were maintained and, in fact, improved by using this glass in conjunction with a silicone binder.[5]

Viscosity and forming temperature of the initial T glasses were reduced by replacing part of the alkaline earths with iron oxide. This glass contained 5 to 12% iron oxide, and viscosity and liquidus were maintained at a low level by proper control of the FeO/Fe_2O_3 ratio. This glass was developed[6] for the production of fine fibers, using the J. M. spinning process. The low forming temperature (925°–1095°C) allows the use of base metal to form the spinning apparatus.

The introduction of the rotary process in the mid-1950s dictated drastic changes in glass composition to reduce viscosity and liquidus while maintaining good water durability. Such compositions are covered in the Welsch patent[7] shown in Table IV. One of the several typical compositions listed in the patent is also given. Viscosity was reduced by increasing the B_2O_3 and alkali levels. The increase in boron also contributed to improved durability. This patent also covers a series of options: (1) addition of MnO, ZnO, and Fe_2O_3, to improve durability and reduce viscosity and (2) addition of TiO_2 and ZrO_2 to improve durability, liquidus, and viscosity. The high alkali content of these glasses also improves melting behavior.

The Welsch patent is very broad, and most of the glasses developed for the rotary process are either included in this patent or are extensions of the initial patent. The main modifications involve reducing B_2O_3 to below 5% to reduce cost and adding fluorine, to maintain or improve durability and liquidus, further reduce viscosity, and improve melting behavior,[8] and/or TiO_2 and ZrO_2[9-11] to maintain or improve properties. Other modifications involve a large increase in ZnO and a reduction in alkali level to improve durability.[12] Fibers with improved durability and lower viscosity were obtained by keeping boron and alkali levels high and adding ZnO.[13]

Continuous Fibers

Three glass types provide the predominant materials for production of textile-grade, continuous glass fiber: E glass (electrical), C glass (chemical), and S glass (high tensile strength).

E Glass: E glass was first formulated in the 1930s, when the need arose for a continuous fiber for insulating fine wires at high temperatures. Due to its excellent mechanical properties and water durability, use of E glass has spread, and over 90% of the continuous glass fiber produced today is comprised of E glass.

E glasses are lime-alumina-borosilicate compositions developed to have both high bulk electrical resistivity and high surface resistivity, as well as good fiber-forming characteristics. "E" glass is based on the eutectic in the system CaO-Al_2O_3-SiO_2, which occurs at 62.2% SiO_2, 14.5% Al_2O_3, 23.3% CaO.

High surface resistivity is obtained by restricting the total alkali content to less than a few percent by weight.

"E" glass need not be a single, definite composition, but may be many compositions within a general range, as given in Table V. The composition may be tailored by each producer to the economics of his raw-material supplies and the details of his production processes. Small changes in composition within the ranges cited do not greatly affect the properties important for E-glass applications.

The original E-glass patent[14] from Dr. Schoenlaub was issued to OCF in 1943. The range of composition covered and a preferred composition are given in Table

Table V. E Glass Composition, Wt%

	Compositional Range	Schoenlaub (Ref. 14) 1940		Tiede/Tooley (Ref. 15) 1948	
SiO_2	52–56	52–56	54	52–56	54
Al_2O_3	12–16	12–16	14	12–16	14
Fe_2O_3	0–0.8				
CaO	16–25	16–19	17.5	19–25	22
MgO	0–6	3–6	4.5		
B_2O_3	5–10	9–11*	10*	8–13*	10*
$Na_2O + K_2O$	0–2				
TiO_2	0–1.5				
F_2	0–1		0–1.25		
T_3 (°C)			1200		1200
Liquidus (°C)			1120		1065

*1 to 3% of fluorspar and 1 to 3% alkali may be substituted for part of the B_2O_3.

V. The low alkali was also claimed to improve water durability. The patent revealed the use of fluorides and sulfates as fining agents and stated that 1–3% alkali and 1–3% fluorspar could be substituted for B_2O_3, but did not cover these modifications in the claims.

The composition limits covered by the patent were relatively narrow, especially those for B_2O_3. Most of the glass that has been made has had B_2O_3 levels below the claimed limits. When the patent was applied for, melting capabilities suggested that the limits given would be adhered to. It seemed that glass with more than 11% B_2O_3 would be too expensive, and with less than 9% would be too difficult to melt. These glasses can be fiberized at around 1200°C.

A composition was later developed in which most of the MgO was substituted by CaO. The range of composition covered in this patent by Tiede and Tooley[15] is given in Table V. Electrical properties, water durability, and viscosity are comparable to those of MgO-containing glasses. The low magnesia level eliminated the need for an additional raw material, such as dolomite, in the batch formulation. A major improvement of this composition is a significant reduction in liquidus. The dissolution rate of zircon and zirconia at furnace operating temperatures is slower. This means that the liquidus is less affected by reaction of the glass with furnace refractories. Furthermore, the devitrification rate is approximately one-half the usual rate with magnesia removed. The reduced crystallization rate results in a glass that can be fiberized with less risk of impaired process efficiency or product quality.

Even the expanded limits claimed in the patent covering low-MgO E glass proved to be too narrow, and perhaps most of the low-MgO E glass that has been made had less B_2O_3 than the minimum claimed in the patent. It should be noted, however, that the patent covered glass with less than 8% B_2O_3, to the extent that fluorospar or alkali were present. The most expensive component in E glass is B_2O_3, and all glass-fiber manufacturers have tried to minimize its use. From a finished-glass property standpoint, there is little penalty in doing so; boric oxide was mainly added to aid melting and reduce forming temperatures. Better refractories, better tank design, more knowledge about batch formulations and melting, and better bushing materials and designs have made it possible to gradually reduce the amount of boric oxide required.

Magnesium oxide and CaO are essentially interchangeable, on a weight-percent basis, in their effect on viscosity in the E-glass system. Substituting MgO for CaO may increase viscosity, but only slightly. The fact that these two oxides are so nearly interchangeable on a *weight-percent basis* suggests that they play a somewhat different role in the glass structure. If they were equivalent structurally, they should be equivalent on a *mole-percent basis,* and a smaller weight of MgO should be required to replace a given amount of CaO. It may be hypothesized that the small size and relatively high charge of the magnesium ion permits it to participate in the glass network.

Further reductions in B_2O_3 are covered by a 1979 patent by Neely.[16]

The value of fluorides as fining agents or melting aids for E glass was recognized almost from the beginning, and is mentioned in the Schoenlaub patent. Fluorides have been used with few exceptions since the 1940s.

The value of sulfates and nitrates was also known early, and these materials have also been used for a long time.

Recent trends in E-glass composition are dictated by the desire of reducing batch cost without degradation of melting, forming, and product performances. Efforts are aimed at further reductions in B_2O_3, since this oxide is still, by far, the most expensive of the batch constituents.

Various modifications of the E-glass composition have been patented. Several approaches have been used to eliminate boron. One approach is to use a composition close to the E-glass eutectic in the system $CaO-Al_2O_3-SiO_2$ and possibly MgO. Such a composition is covered in Slayter's patent.[17] High tensile strength, good chemical durability, and low batch cost are gained at the expense of increased viscosity. Another approach for producing boron- and fluorine-free E-type glasses with forming viscosity comparable to E glass was adding Li_2O and TiO_2 to reduce viscosity and ZnO, SrO, and BaO to reduce the liquidus. These additives have the extra advantage of drastically improving acid durability. Such a composition is disclosed in an Erickson and Wolf patent.[18] It must be pointed out that the compositions listed above are not E-glass compositions.

Standard soda-lime-silicate glass, or A glass, has been used for reinforcement.[1] This glass has several disadvantages that limit its use: (1) high electrical conductivity, which makes the glass improper for electrical applications, and (2) significant strength degradation with time, due to stress corrosion.

The popularity of E glass as an all-purpose glass for continuous fibers makes any change in composition difficult, since manufacturers are not always aware of the needs of their customers and, in fact, in some cases the customers are not aware of all the properties that can affect their finished product.

An interesting application of E glass is as a raw material to prepare high-silica fibers. The E glass has excellent resistance to corrosion by water and moderate resistance to alkali, but is readily attacked by strong mineral acids. In fact, these mineral-acid solutions can, with increased temperature, leach everything out of E-glass fibers but the silica network, even without prior heat treatment. This leaching process, plus a heat treatment, is used commercially to produce high-silica fibers for high-temperature insulation and for reinforcement in ablative composites.

S Glass: S glass was developed as a commercially fiberizable glass with higher strength and elastic moduli than E glass. A comparison of S glass[19] and E glass is shown in Table VI. Basically, there is about a 20% improvement in tensile strength and 15% in elastic moduli. The S glass has a slightly lower density, which is important for mechanical properties considered on a specific basis. Due to the

Table VI. Composition of Glasses Used in Fiber Manufacture and Their Basic Properties

Constituent or Physical Property	E Glass (wt%)	S Glass (wt%)	S Glass (mol%)
SiO_2	52–56	65	69
Al_2O_3	12–16	25	15.5
B_2O_3	5–10		
MgO	0–5	10	15.5
CaO	16–25		
Na_2O	0–2		
K_2O	0–2		
Fe_2O_3	0–0.8		
F_2	0–1		
Tensile strength of single fiber (MPa)			
at 22°C	3800	4500	
at 370°C	2600		
at 540°C	1700		
Tensile strenth of strand (MPa)	2400	3100	
Young's modulus of fiber at 25°C (MPa)	75×10^3	85×10^3	
Density (g/cm³)	2.52–2.62	2.47–2.49	
Refractive index	1.546–1.560	1.524–1.528	
Coefficient of linear thermal expansion per °C	5×10^{-6}	3×10^{-6}	

absence of fluxes, this glass is quite refractory and must be fiberized between 1500° and 1600°C.

Extensive research has been conducted on increasing tensile strength and modulus of glass fibers. The solutions include (1) addition of BeO,[20,21] which markedly raises modulus, although the toxic properties of this material have prevented its commercial use, and (2) addition of Group II oxides, such as cerium oxide, vanadium oxide, or lanthanum oxide.

C Glass: C glass was developed for applications requiring greater corrosion resistance to acids than that of E glass: It has good resistance to most acids. A typical composition is listed in Table VII. This composition is covered by the very broad Bowes patent[2] that covers the steam-blown-wool compositions. The improvement in acid durability over E glass is obtained by increasing silica content and decreasing alumina, while keeping the alkali level moderate.

New Glass Development

The role of many oxides on glass properties is well understood, and tables or charts showing their effects have been published (see, for example, Ref. 22). A summary of such effects is shown in Table VIII. These data can be explained from our knowledge of glass structure and how various oxides enter the structure. This

Table VII. Composition of Acid-Resistant Glasses (C Glasses)

	%
SiO_2	65
Al_2O_3	4
CaO	14
MgO	3
B_2O_3	5.5
$Na_2O + K_2O$	8.5

information, however, is not complete and may apply only to a narrow range of composition. For example, addition of sodium increases or decreases the refractive index, depending on the composition of the glass. This knowledge, combined with the study (or development) of phase diagrams, assists glass scientists in developing new glasses.

The advance of the computer has allowed the glass scientists to considerably speed up the process of computing new glasses or optimizing existing compositions. Using the large files of glass compositions and properties available at most glass companies, it was possible to replace the trends listed in Table VIII by numbers, i.e.,

$$\text{Property} = \sum_i a_i x_i \tag{1}$$

where x_i is the weight or mole percent of the oxide i and a_i is a coefficient specific to this property and oxide. Linear or more complex regressions have been used.

Table VIII. Effect of Various Oxides on Glass Properties

	Strength	Modulus	Density	Refractive Index	Dielectric Constant	Resistivity	Durability Water	Durability Acid	Durability Base
Glass Formers									
SiO_2	+ + +	−	− −	− − −	− − −	+ +	+	+ +	− −
B_2O_3	−	− −	−	−	−	0	+ +	− −	
Intermediates									
TiO_2	0	+	+	+ + +	+		+	+ +	
Al_2O_3	+ +	+	+ +	0	+	+	+	− −	
ZrO_2	0	+	+	+				+	+ + +
ZnO	0	+	+	+ +	+		+	+ +	+ +
BeO	+	+ + +	− −						
Modifiers									
MgO	+	+	+	+	0	0	+	0	
CaO	0	0	+	+ +	0	0	0	− −	+ +
SrO		−	+	+ + +					
BaO	0	0	+ +	+ +	+	0	+	+	
Li_2O	−	0	0	0	+ +	− −	−	−	
Na_2O	− −	− −	+	− −	+ +	− −	− −	− −	+ +
K_2O	− −	− −	+		+ +	− −	− −	−	
F_2	−		−	− −			+	+	

NOTE: A + indicates that addition of the oxide increases the property; + + + indicates a very strong positive effect.

A linear regression is valid for a narrow range of compositions only, but allows simpler computations. Having developed a set of coefficients for a family of glasses either by using existing files or melting series of glasses, it is straightforward to solve the inverse problem of finding a glass composition with given properties. To limit the number of solutions, one extreme condition is normally added, such as lowest cost or lowest liquidus.

Glass compositions for glass fibers with excellent properties and cost have been developed over the years. The trends in development of new glasses takes advantage of new developments in raw materials, melting technology, refractories, forming technology, and metallurgy to either lower the cost of or improve existing glasses, while keeping comparable properties. A major thrust has been to reduce boron content, since it is the most expensive raw material.

References

[1]K. L. Loewenstein, "The E-glass Fibre Industry—An Unneccessarily Restrictive Technology?" *Glass Ind.*, **57** [3] 61 (1980).

[2]V. E. Bowes, "Sodium Calcium Borosilicate Glass," U.S. Pat. No. 2 308 857, December 20, 1939.

[3]D. P. Dingledy, "Glass Composition," U.S. Pat. No. 2 664 359, June 1, 1951.

[4]R. L. Tiede, "Glass Fiber Products," U.S. Pat. No. 3 253 948, February 12, 1962.

[5]J. Stalego, U.S. Pat. No. 2 990 307, June 1961.

[6]W. P. Hahn and E. R. Powell, "Glass Composition," U.S. Pat. No. 2 756 158, September 9, 1952.

[7]W. W. Welsch, "Glass Composition," U.S. Pat. No. 2 877 124, September 25, 1955.

[8]W. W. Welsch, "Glass Composition," U.S. Pat. No. 2 882 173, June 20, 1955.

[9]S. deLajarte, "Glass Composition," U.S. Pat. No. 3 013 888, September 6, 1959.

[10]B. Laurent and C. Haslay, "Fiberizable Glass Composition," U.S. Pat. No. 3 508 939, January 14, 1965.

[11]B. Laurent and C. Haslay, "Silicate Glass Compositions," U.S. Pat. No. 3 853 569, July 1, 1970.

[12]C. Haslay, Agnetz, and J. Paymal, "Manufacture of Borosilicate Fibers," U.S. Pat. No. 3 523 803, August 11, 1970.

[13]L. V. Gagin, "Glass Composition for Fiberization," U.S. Pat. No. 4 177 077, September 19, 1978.

[14]R. A. Schoenlaub, "Glass Composition," U.S. Pat. No. 2 334 961, December 5, 1940.

[15]R. L. Tiede and F. V. Tooley, "Glass Composition," U.S. Pat. No. 2 571 074, November 2, 1948.

[16]H. E. Neely, Jr., "Glass Fiber Composition," U.S. Pat. No. 4 166 747, October 13, 1977.

[17]G. Slayter, U.S. Pat. No. 2 309 039, 1945.

[18]T. D. Erickson and W. W. Wolf, "Glass Composition Fibers and Methods of Making Same," U.S. Pat. No. 4 026 715, August 22, 1984.

[19]R. S. Harris and G. R. Machlan, "Glass Composition," U.S. Pat. No. 3 402 055, July 12, 1965.

[20]G. L. Thomas, "High Tensile Strength, Low Density Glass Composition," U.S. Pat. No. 3 183 104, October 26, 1962.

[21]R. M. McMarlin, "Glass Composition," U.S. Pat. No. 3 620 787, October 2, 1968.

[22]The Handbook of Glass Manufacture. Compiled and edited by F. V. Tooley. Books for the Glass Industry Div., Ashlee Books, Inc., New York, 1984.

Glazes and Enamels

RICHARD A. EPPLER

Consultant
400 Cedar Lane
Cheshire, CT   06410

Glazes and enamels are vitreous coatings on ceramics and metals, respectively. These coatings may render the substrate impervious, mechanically stronger, more resistant to abrasion, chemically more inert, more readily cleanable, and esthetically more pleasing to the eye. The formulations and properties of these materials will be discussed in this paper.

Glass is used not only as a monolithic material, but also as a coating; a glass coating applied to a ceramic substrate is referred to as a *glaze,* and a vitreous coating applied to a metal substrate as a *porcelain enamel.* In either case, the coatings are thin layers of glass that have been fused onto the surface of the substrates.

Glass coatings are applied to substrates for several reasons[1]: In any given example, a glass coating may make the substrate chemically more inert and impervious, more readily cleanable, more resistant to abrasion and scratching, mechanically stronger, and esthetically more pleasing to the touch and eye.

Some of the physical requirements for a glass-coating material relate to its use.[1] In almost all cases, a coating must have a homogeneous, smooth, and hard surface to resist abrasion and scratching and to facilitate cleaning. An exception is a textured coating, where an esthetically pleasing pattern is deliberately imposed. A smooth surface is not only visually more appealing and resistant to abrasion and scratching, but also more likely to be impervious to liquids and gases and thus more readily cleanable. Many applications of glass coatings involve contact with food and drink: For such applications, high standards of cleanability are essential.[2] Moreover, for many if not for most glass coatings, chemical durability in service is a prime reason for selection of the vitreous coating material. Glass coatings are formulated for resistance to many reagents, including hot water, acids, alkalies, and most if not all organic media. Practically the only reagent to which these materials cannot be exposed is hydrofluoric acid.

A coating material, however, also has requirements deriving from the fact that it must be applied to and bond with the substrate.[3] It must fuse to a homogeneous, viscous glass at a temperature either coincident with that at which the substrate matures or low enough that distortion of the substrate during firing of the coating does not occur.

As the coating matures, it must react with the surface of the substrate to form an intermediate bonding layer.[4] In some porcelain enamels, components of the coating, called adherence oxides, may be included to produce this reaction. In all applications, the formulation of the coating must allow just the right amount of interaction with the body during firing. If the reaction is insufficient, the coating will not bond and, therefore, will fall off the substrate. On the other hand, if the reaction is excessive, the composition of the coating or the body may be affected adversely, particularly through the introduction of defects.

The coefficients of expansion of the coating and the substrate must be related to each other so that excessive stresses and strains do not result from cooling the fired ware, causing defects such as spalling or crazing.[5] Although the expansions of the coating and the substrate should be close, they should not be identical: Glass coatings and ceramic bodies, unlike metals, are much stronger in compression than in tension. For optimum performance, therefore, the coating, which is thinner, is compressed. This is done by selecting a coating with a coefficient of thermal expansion somewhat lower than that of the substrate; thus, during cooling the coating shrinks less than the substrate. The coating must also have a low surface tension, so that it will spread uniformly over the substrate and not crawl away from edges and holes.

Ceramic Glazes

The vitreous coatings applied to ceramic ware are called *ceramic glazes*. So many glaze formulations are used that it is impossible to classify glazes simply.[1] A primary reason for this variety of glaze compositions is that ceramic ware is fired over a very wide temperature range ($\approx 800°$ to $>1400°C$).[6] Obviously, the same glaze composition would not be satisfactory at all temperatures. Any given composition is useful for a temperature range of only $\approx 30°C$. Thus, recipes are required for each range of temperature.

Another reason for a variety of glaze compositions is the desire for a variety of surface qualities. Glazes may be bright or dull, glossy or mat, opaque or transparent, thick or thin, as required.

We will begin this discussion of the various types of glazes with leadless glazes. Glazes containing lead will be treated separately because of lead's unique character and because of the many problems resulting from the poisonous nature of lead oxide. Then we will introduce opaque glazes and techniques used for opacification. Finally, the development of satin and mat glazes will be treated.

Leadless Glazes

When lead oxide is not used as a flux, one must rely on basic oxides such as calcium oxide, magnesium oxide, and the alkali oxides, together with boric oxide, for fluxing. Nevertheless, a number of leadless glaze systems have been known for many years.[7] In the first place, glazes that are fired to $>1150°C$ must be leadless because glazes containing lead break down at $\approx 1150°C$, with excessive volatilization of lead oxide. Column 1 of Tables I and II shows the formula of one of these high-temperature leadless glazes.[8] Such glazes are used on the green body of hard-paste porcelain. The glazes designed for these highest-firing porcelains originated from studies of feldspar and the system calcia-alumina-silica.

For porcelains fired at lower temperatures — the soft porcelains or hard stoneware — formulas such as those in column 2 of the tables are satisfactory.[7-9] The overall amount of alumina and silica used is adjusted for the firing temperature required. The usual range is 1 flux, $\frac{1}{2}$–1 alumina, and 6–15 silica for cones 11–16; and 1 flux, 0.4–0.8 alumina, 3–5 silica for cones 7–10. The ratio of silica to alumina is held within the narrow range from 7:1 to 10:1.

A related glaze is that used on sanitary ware,[10] as shown in column 3 of the tables. Sanitary-ware glazes mature at a high enough temperature that limitations imposed by fusibility do not apply; however, since tin oxide, an extremely expensive material, is a customary addition for opacity, the composition must be adjusted to lower the solubility of the opacifier in the glaze. To this end, boric acid

Table I. Selected Commercial Glazes, Mole Ratio*

Oxide Mole Ratio	1 Glaze for Hard-Paste Porcelain	2 Glaze for Soft-Paste Porcelain	3 Sanitary-Ware Glaze	4 Bristol Glaze	5 Wall-Tile Glaze	6 Semivitreous Dinnerware Glaze	7 Vitreous Dinnerware Glaze
Li_2O						0.047	0.070
Na_2O	0.3	0.3	0.10	0.1	0.27	0.081	0.069
K_2O			0.10	0.1	0.04	0.115	0.037
MgO				0.2	0.01	0.066	0.391
CaO	0.7	0.7	0.60	0.4	0.35	0.580	0.322
ZnO			0.20	0.2	0.32		0.071
SrO							0.039
BaO						0.110	
PbO							
CoO							
NiO							
CuO							
SiO_2	10.0	7.0	3.00	3.5	2.65	2.721	2.224
Al_2O_3	1.0	0.8	0.55	0.4	0.26	0.367	0.173
B_2O_3					0.05	0.171	0.188
ZrO_2							0.011
P_2O_5							
TiO_2							
MnO_2							
Cr_2O_3							
Sb_2O_3							
Nb_2O_5							
WO_3							
MoO_3							
F							

Table I. (continued)

Oxide Mole Ratio	8 Low-Expansion Semicrystalline Glaze	9 Dinnerware Glaze	10 Cone 06 Dinnerware Glaze	11 Opacified Glaze	12 Zinc-Mat Glaze	13 Lime-Mat	14 Ground-Coat Enamel	15 Home-Laundry Enamel
Li_2O	0.85							0.085
Na_2O		0.179	0.157	0.111	0.087	0.040	0.065	0.638
K_2O	0.15	0.066		0.083	0.052	0.059	0.470	0.052
MgO					0.097		0.054	0.014
CaO		0.494	0.218	0.407	0.152	0.524	0.244	0.157
ZnO				0.371	0.364			0.010
SrO								
BaO		0.261		0.028	0.247	0.377	0.105	0.015
PbO			0.625					
CoO							0.014	0.015
NiO							0.038	0.013
CuO							0.008	
SiO_2	2.51	3.369	2.792	2.019	1.566	1.746	1.623	2.171
Al_2O_3	0.74	0.340	0.273	0.406	0.443	0.262	0.138	0.354
B_2O_3	0.20	0.314	0.507	0.143	0.142	0.176	0.489	0.721
ZrO_2			0.023	0.248	0.210	0.211		0.162
P_2O_5							0.011	0.010
TiO_2								0.100
MnO_2							0.005	0.024
Cr_2O_3								
Sb_2O_3								
Nb_2O_5								
WO_3								
MoO_3								
F							0.316	0.381

Table I. *(continued)*

Oxide Mole Ratio	16 Hot-Water-Tank Enamel	17 Continuous-Clean Coating	18 Opaque Cover-Coat Enamel	19 Semiopaque Cover-Coat Enamel	20 Clear Cover-Coat Enamel	21 Architectural Cover Coat
Li_2O	0.135	0.051	0.127	0.129	0.198	0.220
Na_2O	0.679	0.347	0.646	0.486	0.665	0.680
K_2O		0.046	0.277	0.341	0.137	
MgO						
CaO	0.110	0.034				0.007
ZnO	0.047			0.045		0.165
SrO						
BaO	0.011					
PbO						
CoO	0.019	0.001				
NiO		0.001				
CuO		0.518				
SiO_2	2.820	1.186	2.902	2.730	3.313	2.026
Al_2O_3	0.060	1.195	0.094	0.046	0.090	0.044
B_2O_3	0.330	0.050	0.986	0.833	0.344	0.402
ZrO_2	0.286	0.173			0.215	0.070
P_2O_5			0.039			0.011
TiO_2			1.117	0.582	0.151	0.807
MnO_2	0.063	0.001				
Cr_2O_3		0.024				
Sb_2O_3		0.003				
Nb_2O_5			0.001			0.008
WO_3			0.001			
MoO_3					0.011	
F	0.349	0.112	0.710	0.724	0.417	0.082

*The glaze compositions documented in this table are expressed in mole ratio, as is the convention of the commercial-glaze industry. The sum of the modifier oxides (top portion of the table) is normalized to 1, and the glass-forming oxides are expressed relative to that sum.

Table II. Selected Commercial Glazes, Wt Pct*

Oxide Wt%	1 Glaze for Hard-Paste Porcelain	2 Glaze for Soft-Paste Porcelain	3 Sanitary-Ware Glaze	4 Bristol Glaze	5 Wall-Tile Glaze	6 Semivitreous Dinnerware Glaze	7 Vitreous Dinnerware Glaze
SiO_2	78.00	73.83	59.71	67.09	62.25	59.09	55.79
Al_2O_3	13.24	14.32	18.58	13.01	10.36	13.53	7.37
B_2O_3							
ZrO_2					1.36	4.30	5.47
P_2O_5							0.57
TiO_2							
Li_2O						0.51	
Na_2O	3.67		2.05	1.98	6.54	1.81	1.81
K_2O		4.96	3.12	3.01	1.47	3.92	2.71
MgO				2.57	0.16	0.96	0.62
CaO	5.10	6.89	11.15	7.16	7.67	11.76	9.16
ZnO			5.39	5.19	10.18		10.94
SrO						4.12	3.07
BaO							2.50
PbO							
CoO							
NiO							
CuO							
MnO_2							
Cr_2O_3							
Sb_2O_3							
Nb_2O_5							
WO_3							
MoO_3							
F							

Table II. (continued)

Oxide Wt%	8 Low-Expansion Semicrystalline Glaze	9 Dinnerware Glaze	10 Cone 06 Dinnerware Glaze	11 Opacified Glaze	12 Zinc-Mat Glaze	13 Lime-Mat	14 Ground-Coat Enamel	15 Home-Laundry Enamel
SiO_2	53.91	55.88	42.45	44.07	33.31	35.99	44.01	41.55
Al_2O_3	26.98	9.57	7.04	15.04	15.99	9.17	6.35	11.50
B_2O_3	4.98	6.04	8.93	3.62	3.50	4.20	15.37	15.99
ZrO_2			0.72	11.10	9.16	8.92		6.36
P_2O_5							0.70	0.45
TiO_2								2.55
Li_2O	9.08							0.81
Na_2O		3.06	2.46	2.50	1.91	0.85	0.88	12.60
K_2O	5.05	1.72		2.84	1.73	1.91	13.15	1.56
MgO					1.38		2.30	0.18
CaO		7.65	3.09	8.29	3.02	10.08	6.18	2.80
ZnO				10.97	10.48			0.26
SrO								
BaO				1.56		28.87	7.27	0.73
PbO		16.08	35.30		19.52			
CoO							0.47	0.36
NiO							1.28	0.31
CuO							0.29	
MnO_2							0.20	0.66
Cr_2O_3								
Sb_2O_3								
Nb_2O_5								
WO_3								
MoO_3								
F							2.71	2.31

Table II. *(continued)*

Oxide Wt%	16 Hot-Water-Tank Enamel	17 Continuous-Clean Coating	18 Opaque Cover-Coat Enamel	19 Semiopaque Cover-Coat Enamel	20 Clear Cover-Coat Enamel	21 Architectural Cover Coat
SiO_2	56.05	24.20	40.97	46.74	59.07	37.61
Al_2O_3	2.02	41.38	2.25	1.34	2.72	1.39
B_2O_3	7.60	1.18	16.13	16.53	7.11	8.65
ZrO_2	11.66	7.24			7.86	2.67
P_2O_5			1.30			0.48
TiO_2		0.03	20.97	13.25	3.58	19.93
Li_2O	1.33	0.52	0.89	1.10	1.76	
Na_2O	13.92	7.30	9.41	8.58	12.23	4.21
K_2O		1.47	6.13	9.15	3.83	19.79
MgO						
CaO	2.04	0.65				0.12
ZnO	1.27			1.04		4.15
SrO						
BaO	0.56					
PbO						
CoO	0.47	0.03				
NiO		0.03				
CuO		13.99				
MnO_2	1.81	0.03				
Cr_2O_3		1.24	0.06			
Sb_2O_3		0.30	0.05			0.72
Nb_2O_5						
WO_3						
MoO_3					0.47	
F	2.19	0.72	3.17	3.92	2.35	0.48

*The glaze compositions documented in this table are identical to the compositions in Table I. They are expressed in weight percent rather than mole ratio.

is normally omitted and a substantial concentration of barium oxide or zinc oxide added.

The Bristol glaze,[8] shown in column 4, is a variation of the soft-porcelain glaze developed to produce an opaque white coating on stoneware and other colored clay bodies. The opacity arises from high concentrations of zinc oxide. A Bristol glaze can be formulated from porcelain glazes by substituting large quantities of zinc oxide for the alkalies and alkaline earths.

Further modification of this glaze yields suitable material for the glazing of wall tile at rapid firing rates[11]: An example of this type of glaze is given in column 5. The primary alteration consists of reducing silica content to lower firing temperature. In addition, boric oxide is added in low concentrations to further enhance the melting rate. This glaze is suitable for firing in a kiln with a 1–4-hour total cycle.

The development of glazes for semivitreous and vitreous dinnerware is more difficult. Among the required properties are a firing temperature of approximately cone 4, compatibility with essentially all pigment systems stable at cone 4, and a coefficient of thermal expansion no greater than $7.0 \times 10^{-6}/°C$ for semivitreous ware and $5.5 \times 10^{-6}/°C$ for vitreous hotel china.

Column 6 of the tables shows a formulation developed for semivitreous ware.[12] Note the use of more than one alkaline earth: A glaze containing several alkaline earths has been found superior in melting and surface to a glaze containing any of these materials alone, in larger concentrations.[7]

The development of leadless glazes for vitreous dinnerware having even lower expansion imposes more stringent requirements because of the lower coefficient of thermal expansion[13]; this procedure thus has been attempted only recently. An example of such a glaze is shown in column 7. Its unique aspect is the high ZnO content. Zinc oxide has traditionally been unsuitable for use in glazes where a full palette of colors is required, because in most glaze systems, ZnO profoundly affects several ceramic colors. It has just recently been found that ZnO can be used within a small range of glass composition without unacceptable pigment degradation.

The low-expansion glazes suitable for zircon and cordierite bodies require a unique glaze.[14] It is not possible to glaze these bodies with fully vitreous glazes. It is, however, possible to glaze cordierite by including in an appropriately formulated glaze the precipitation of a low-expansion phase, as shown in column 8 of the tables.

Lead-Containing Glazes

Lead oxide is used in glazes for several reasons.[15] First, the strong fluxing action of lead oxide allows the formulation of glazes that mature at lower temperatures than do their lead-free counterparts. Moreover, this fluxing action leads to greater flexibility in the formulation of the glaze for obtaining other desired properties such as low expansion and smooth surface. Second, lead oxide in the glaze allows for satisfactory maturing of the glaze over a wider firing range; thus, leaded glazes are more adaptable to the varying conditions that occur in production-scale equipment. Moreover, lead oxide imparts low surface tension, leading to a smooth, high-gloss surface. Lead-containing glazes also heal over defects in the glaze surface more readily. Furthermore, lead oxide imparts to a glaze a high index of refraction, which results in a brilliant appearance. Finally, lead oxide in the glaze reduces surface crystallization or devitrification of the glaze. This particular combination of desirable properties is difficult to achieve on a production scale in leadless glazes.

On the other hand, there are several disadvantages to the use of lead oxide in glazes. In the first place, lead-glazed ware must be fired in a strongly oxidizing atmosphere, as lead is very readily reduced. Lead oxide also volatilizes above ≈1200°C: Hence, lead glazes are seldom used above cone 6. By far the most serious disadvantage is that lead oxide is poisonous. Appropriate precautions must be taken in using it, to avoid even the possibility of lead poisoning. Lead poisoning is caused by the ingestion of soluble lead compounds into the system, usually by mouth, although it can also result from breathing vapors or dust. The disease is very difficult to diagnose, since its various symptoms are similar to those of many other ailments. When preparing lead glazes, therefore, every possible precaution must be taken to avoid poisoning.

Moreover, if glazes are not formulated properly, they are subject to attack, primarily by acidic media, which results in release of lead into the solution. If such glazes are used in contact with food or drink, lead poisoning may result.

An example of a lead-containing dinnerware glaze is shown in column 9 of Tables I and II.[16] Compared with leadless glazes for similar applications, the silica, boric oxide, alumina, and calcia contents are similar. Using lead oxide in the glaze permits use of higher concentrations of alkali oxide, resulting in greater fluidity and improved surface.

To further reduce the maturing temperature, alumina and silica are decreased and lead-oxide content increased. An example of a commercial clear-glaze formula suitable for artware and hobbyware bodies at cone 06 is shown in column 10.[17]

The principal limitation of lead oxide in glazes is its potential for lead poisoning. Unfortunately, over the years occasional episodes of lead toxication have resulted from using improperly fired and formulated lead-containing glazes on ceramic ware.[18] This problem has been studied, and a test for lead release that is suitable for the screening of formulations has been developed. The test is applicable to ceramic glazes ranging from those with negligible lead release to those with high release. It is the basis for federal regulations, applicable to food-contact surfaces, that establish limits for lead release from any ceramic glaze.

Opaque Glazes

Opaque glazes transmit low enough light through the coating that they effectively hide the body. Whiteness or opacity is introduced into glass coatings by adding a substance that disperses into the coating as discrete particles, scattering and reflecting some of the incident light.[10] To do this, the dispersed substance must have a low solubility in the molten glaze and a refractive index that differs appreciably from that of the clear glass coating. The refractive index of most ordinary glasses is 1.5–1.6 and that of opacifiers must be either greater or less. Practically, opacifiers of high refractive index are used. Some possibilities include tin oxide, with a refractive index of 2.04; zirconia, with a refractive index of 2.40; zircon, with a refractive index of 1.85; and titania, with a refractive index of 2.5 for anatase or 2.7 for rutile.

As shown in column 11 of Tables I and II, for glazes fired at >1000°C, zircon is the best opacifier.[19] It has a solubility in many ceramic glazes of about 5% at 1200°C and 2–3% at 700°C. A customary total addition would be 8–10% zircon. Hence, most opacified glazes contain both zircon that was placed in the mill and went through the firing process unchanged and zircon that dissolved in the molten glaze during firing but recrystallized with cooling. In coatings where the firing temperature is considerably less than 1000°C, titania in the anatase crystal phase is the best opacifying agent because of its very high index of refraction. At 850°C, however, anatase inverts to rutile in silicate systems. Once inverted to rutile, titania crystals can grow rapidly to sizes that are no longer effective for opacification. Moreover, because the absorption edge of rutile is very close to the visible spectrum, it extends into the visible as the rutile particles grow, leading to a pronounced cream color.

Satin and Mat Glazes

Satin and mat effects[10] are also due to small crystals dispersed in the glaze, the result of devitrification when a completely fused glaze cools and part of the fused mass crystallizes. The crystals must be very small and evenly dispersed to give the glaze surface a smooth and velvet appearance. Mat glazes are always somewhat opaque because the crystals, as in normal opaque glazes, break up light rays. Crystals are zinc silicate in the case of zinc mats (column 12) or calcium silicate in the case of lime mats (column 13).

Porcelain Enamels

A glass coating applied to a metal substrate is called a porcelain enamel. The interface between the glass coating and the metal substrate is critical for bonding between the two materials.[4] The glass coating must act not only as a protective and esthetically pleasing surface, but also must effectively bond to the substrate metal. Therefore, adherence between enamels and metal substrates and the associated requirements imposed on the coating will be discussed before proceeding to porcelain-enamel materials.

For proper adherence, the enamel at the interface must be saturated with an oxide of the base metal.[4] This oxide, which for iron and steel substrates is FeO, must not be reduced by the metal when in solution in the glass. Furthermore, the interfacial region must be composed so that a continuous electronic structure or chemical bond exists in the interfacial zone.[20] On the other hand, surface roughness has been found of little value when the chemical bond is weak, although a rough surface generally improves adherence. A transition zone must be in equilibrium and compatible with both the metal and the glass coating at the interface to create a continuity between the atomic and electronic structures of the metal and the ceramic. This zone must include at least a monomolecular layer of the oxide of the metal substrate and is stable as long as both the metal and the glass coating at the interface are saturated with the metal oxide. If a substrate is heated in air, a scale forms on it before the coating sinters to a continuous mass.[4] After the enamel fuses, it attacks and dissolves the scale. This is necessary because such oxide layers lack mechanical strength, resulting in flaking-off of the coating.[21]

Dissolution of the oxide scale causes contact between the glass and the surface of the metal. A chemical reaction occurs which oxidizes the substrate surface until both the substrate and the glass coating at the interface are saturated with the metal oxide.

Adding certain ions, such as cobalt oxide, nickel oxide, and copper oxide, to the ceramic-coating formulation results in improved adherence between the glass coating and the metal.[4] These oxides contribute substantially to the rate at which saturation of the substrate and the coating with the oxide of the substrate occurs. They also play a critical role in creating and maintaining the saturation at the interface between the substrate and the coating.

Column 14, Tables I and II, shows a general-purpose ground-coat enamel. These materials are basically alkali borosilicate formulations containing small amounts of adherence-promoting oxides such as cobalt, nickel, manganese, and copper.[22] These enamels are formulated to have a satisfactory bond or adherence to the base metal. They also provide a chemically resistant protective layer, which minimizes surface defects that may be caused by either the substrate itself or its preparation method.

Ground-coat frits can also be formulated for particular end-use purposes. For example, in column 15 is an enamel formulated for outstanding alkali resistance by adding large quantities of zirconium oxide.[23] These enamels are used in the home-laundry industry. When the outstanding thermal and corrosion resistance required of hot-water-tank systems are needed, an enamel such as shown in column 16 is used.[23] The higher concentration of silica and lower concentration of boron oxide in these coatings reflect the substantially higher firing temperature accepted by the hot-water-tank industry for improved chemical and thermal resistance.

The continuous-clean coating is a substantially different type of ground-coat formulation.[24] This coating has been developed for volatilizing and removing food soils from the internal surfaces of ovens at normal operating temperatures; an example is shown in column 17. In these materials, various active ingredients promote volatilization, and the alumina-to-silica ratio is increased to provide limited porosity.

Cover-coat porcelain enamels are formulated to provide specific color and appearance characteristics, surface hardness, abrasion resistance, and resistance to heat and thermal shock and to atmospheric and liquid corrosion.[22] They can be opaque, semiopaque, or clear. Opaque enamels are used for white and pastel coatings, semiopaque enamels for most medium-strength colors, and clear enamels for bright, strong colors.

Opaque enamels, as shown in column 18 of the tables, are opacified with titanium dioxide.[25-27] Usually, all of the titanium dioxide is smelted to a clear frit, which partially crystallizes to anatase and rutile during the firing process, providing the required opacification. The materials have excellent acid resistance and fairly good alkali resistance. Titania-opacified enamels are produced with reflectances of 78 to 88%.

As with the other porcelain enamels, the glass is basically an alkali borosilicate. Large concentrations of titanium dioxide are added in the range where it will be soluble at the melting temperature of 1400°C but only partially soluble at the firing temperature of 800°C. Large concentrations of fluoride are added to reduce the stability of the vitreous oxide matrix. Small additions of materials that affect the color of the enamel by altering the crystallization properties are often made[27-29]: These materials include phosphorus pentoxide, niobium pentoxide, and tungsten oxide.

Column 19 gives an example of a semiopaque cover-coat enamel.[22] These materials do not differ qualitatively from the fully opaque enamels; rather, the titanium dioxide concentration is reduced to make the system compatible with the use of pigments for producing colored enamels.

Clear cover-coat porcelain enamels are used in conjunction with appropriate pigments for producing strong and medium-strength colors (an example is shown in column 20).[22] Although some titanium dioxide is present for improving acid resistance, the concentration is low enough that substantial crystallization does not occur. Therefore, inclusion of pigment in the mill formulation permits development of strong colors.

The architectural siding industry needs low-gloss cover-coat enamels with good weatherability and durability. Column 21 of Tables I and II gives an example of an enamel for this application.[22]

Summary

This brief overview has described the variety of ways in which glass can be used as a coating material. Various materials used to form vitreous coatings have been examined. To a great extent, these materials are composed from basic formulations that are varied to meet requirements of suitable firing temperature, adherence, and coefficient of thermal expansion. At very high temperatures and low expansions, simple variations on silica are suitable. As temperature is reduced, ingredients such as alkalies, alkaline earths, lead oxide, and boric oxide are added to provide the necessary firing temperature, durability, and coefficient of thermal expansion. When dissimilar materials are to be joined, special ingredients promoting adherence may be required. The end result is a product in which the many beneficial qualities of glass surfaces are imparted to other substrates.

References

[1] R. A. Eppler, "Glazes and Enamels"; Ch. 4, pp. 301–38 in Glass Science and Technology, Vol. 1, Glass-Forming Systems; Edited by D. R. Uhlmann and N. J. Kreidl. Academic Press, New York, 1983.

[2] J. S. Nordyke, "International Conference on Ceramic Foodware Safety Keynote Address"; pp. 5–8 in Proceedings of the International Conference on Ceramic Foodware Safety, Lead Industries Association, New York, 1974.

[3] F. Singer and S. S. Singer, Industrial Ceramics. Chapman and Hall, London, 1963.

[4] B. W. King, H. W. Tripp, and W. H. Duckworth, "Nature of Adherence of Porcelain Enamels to Metals," J. Am. Ceram. Soc., 42 [10] 504–25 (1959).

[5] W. D. Kingery, H. K. Bowen, and D. R. Uhlmann; p. 609 in Introduction to Ceramics, 2d ed. Wiley & Sons, New York, 1976.

[6] D. Rhodes, Clay and Glazes for the Potter, rev. ed. Chilton, Radnor, PA, 1973.

[7] J. E. Marquis and R. A. Eppler, "Leadless Glazes for Dinnerware," Am. Ceram. Soc. Bull., 53 [5] 443–45, 49; [6] 472 (1974).

[8] P. Rado, An Introduction to the Technology of Pottery. Pergamon, Oxford, England, 1969.

[9] C. W. Parmalee and C. G. Harmon, Ceramic Glazes, 3d ed. Cahners, Boston, 1973.

[10] F. Singer and W. L. German, Ceramic Glazes. Borax Consolidated, Ltd., London, 1964.

[11] W. H. Orth, "Effort of Firing Rate on Physical Properties of Wall Tile," Am. Ceram. Soc. Bull., 46 [9] 841–44 (1967).

[12] Pemco Technical Notebook on Ceramic Glazes and Stains, Pemco Ceramics Group, Mobay Chemical Corp., Baltimore, MD.

[13] R. A. Eppler, L. D. Gill, and E. F. O'Conor, "Recent Developments in Leadless Glazes," Ceram. Eng. Sci. Proc., 5 [11–12] 923–32 (1984).

[14] R. A. Eppler and E. F. O'Conor, "Semicrystalline Glazes for Low Expansion Whiteware Bodies," Am. Ceram. Soc. Bull., 52 [2] 180–84 (1973).

[15] R. A. Eppler, "Formulation and Processing of Ceramic Glazes for Low Lead Release"; pp. 74–96 in Proceedings of the International Conference on Ceramic Foodware Safety, Lead Industries Association, New York, 1974.

[16] J. E. Marquis, "Lead in Glazes—Benefits and Safety Precautions," Am. Ceram. Soc. Bull., 50 [11] 921–23 (1971).

[17] R. A. Eppler, "Formulation of Glazes for Low Pb Release," Am. Ceram. Soc. Bull., 54 [5] 496–99 (1975).

[18] J. S. Cole, "Health Aspects of ILZRO's Lead in Glazes Research," Am. Ceram. Soc. Bull., 50 [11] 917–19 (1971).

[19] F. T. Booth and G. N. Peel, "Principles of Glaze Opacification with Zirconium Silicate," Trans. Br. Ceram. Soc., 58 [9] 532–64 (1959).

[20] J. A. Pask, "Chemical Reaction and Adherence at Glass-Metal Interfaces"; pp. 1–16 in Proceedings of the PEI Technical Forum, Vol. 33, 1971.

[21] B. W. King and B. W. Stull, "One Fire White Enamels," J. Am. Ceram. Soc., 32 [1] 34–40 (1949).

[22] Pemco Porcelain Enamel Technical Manual, Pemco Ceramics Group, Mobay Chemical Corp., Baltimore, MD, 1970.

[23] R. A. Eppler, R. L. Hyde, and H. F. Smalley, "Resistance of Porcelain Enamels to Attack by Aqueous Media: I, Tests for Enamel Resistance and Experimental Results Obtained," Am. Ceram. Soc. Bull., 56 [12] 1064–67 (1977).

[24] P. G. Monteith, O. C. Linhart, and J. S. Slaga, "Performance Tests for Properties of Low Temperature Thermal Cleaning Oven Coatings"; pp. 73–79 in Proceedings of the PEI Technical Forum, Vol. 32, 1970.

[25] R. A. Eppler, "Crystallization and Phase Transformation in TiO_2 Opacified Porcelain Enamels: II," J. Am. Ceram. Soc., 52 [2] 94–99 (1969).

[26] R. F. Patrick, "Some Factors Affecting the Opacity, Color, and Color Stability of Titania-Opacified Enamels," *J. Am. Ceram. Soc.,* **34** [3] 96–102 (1951).

[27] R. D. Shannon and A. L. Freidberg, "Titania-Opacified Porcelain Enamels," *Ill. Univ., Eng. Exp. Sta. Bull.,* **456** (1960); 49 pp.

[28] R. A. Eppler and G. H. Spencer-Strong, "Role of P_2O_5 in TiO_2-Opacified Porcelain Enamels," *J. Am. Ceram. Soc.,* **52** [5] 263–66 (1969).

[29] R. A. Eppler, "Niobium and Tungsten Oxides in Titania-Opacified Porcelain Enamels," *Am. Ceram. Soc. Bull.,* **52** [12] 879–81 (1973).

Sealing Glasses

J. FRANCEL

Consultant
Owens-Illinois, Inc.
Toledo, OH

Commercially used sealing glasses are described and their physical and chemical properties tabulated. The fields of glass-metal and solder sealing glasses are covered in detail.

Commercial sealing glasses join two or more materials to insulate metal conductors, assure vacuum-tight assemblies, and hermetically isolate sensitive electronics components from a harmful environment. These sealing glasses may be divided into groups, according to Table I.

This paper covers the glass–metal and solder–glass groups in more detail. The other groups in the table are also important commercially, but they are already well known: Scientific glassware joints, monochronic television parts, and ceramic glazing are produced by efficient production processes and have been for many years.

Glass–Metal Seals

Glass–metal seals are vacuum-tight joints of glasses and metals used in the electronics industry for electrical insulation and are valuable for their imperviousness to air and gases. More than four decades of experience make this technology very effective in meeting technical, physical, and chemical requirements: Its high reliability ensures long-lived joints at competitive costs. Sealing glass is effective in glass–metal assemblies because of its excellent electrical insulating properties, high impermeability, selective light transmissions, and absorptions.

Materials Selection

Compatibility in thermal expansions is the most basic criterion in selecting glass for specific metal combinations. An "expansion match" of under 100 parts per million in differential contraction is considered safe for all applications. On the other hand, contraction differentials of 1000 parts per million are not recommended for good seals. These contraction differentials are determined by residual stresses developed in cooling the glass–metal joint from setting point to room temperature. Setting point is defined as the temperature at which glass becomes rigid ($=\log$ viscosity 13–14.5 P), usually about 5°C above the strain point. The setting point is not a fixed temperature for a given glass, as are the annealing or strain points, but varies with geometry, thickness, and cooling rate of the glass. Seals of metal and matched glass are appropriately known in the industry as *matched seals*. Table II shows the classification of seals according to expansion considerations, and Table III shows the practical scale of matched seals in relation to elongation differentials. The final intended use of the glass–metal seals guides selection of the seal type.

Table I.   Sealing-Glass Types

Types	Examples
Glass–glass	Television — monochrome
Glass–metal	Feedthrough wire
Glass–ceramic	ROM — window
Glass A–solder glass–glass B	Television — color
Ceramic–glass–metal–glass–ceramic	Cer–DIP

In addition to matched seals, unmatched seals are also used frequently when the sealing glass is in compression. These seals, appropriately named *compression seals,* are quite strong and primarily used in feedthroughs, relay heaters, and sight windows. They can be subjected to severe mechanical and thermal shocks, since the glass compressive strength exceeds its tensile strength by a significant factor of 20. The seals are produced by placing the higher-expansion metal outside of the lower-expansion sealing glass. During the cooling cycle the external metal shrinks around the glass, compressing it. The metal's tensional stresses can be tolerated by relatively thick metal walls.

Seal Selection

Matched and compression seals have been described, but some additional information may be useful: The metal sheet should be ≈ 0.5 mm (≈ 20 mils) thick for a good matched seal. A great variety of designs is possible when meeting the customers' specifications with matched seals, which are light but sensitive to mechanical and thermal stresses and shocks. Compression seals are useful for thick metals. The wall thickness should be greater than 0.5 mm (20 mils) and the height at least 1.5 mm (60 mils), the ratio of the outside to the inside diameter of the metal ring preferably ≈ 1.3, and the glass fillet below the metal line or the glass above this line may crack due to radial tensile strength.

Soft glass can be sealed to the proper metal at relatively low temperatures. For higher thermal shock and corrosion resistance, hard glass may be a better choice than soft glass at low temperatures. Solder glasses fill the need for relatively low-temperature seals on parts which could deform at higher temperatures and also for sealing enclosed devices that are sensitive to high temperatures. Metallized glass is joined to metal by metal solder, without any sealing glass, and adhesive and mechanical seals require no sealing glasses.

Table II.   Classification of Seals by Expansion Considerations

Seal Type	Applications
Matched	Thin metal, caps, disks, shells
Compression	Thick metal, feedthroughs, windows
Soft glass	Low temperature, high expansion
Hard glass	High temperature, low expansion, good thermal shock
Low-melting	Solder glasses, Cer-DIP
Metallized glass	Low-temperature metal solder
Adhesive	Epoxy, silicones, cements, plastic dip
Mechanical	Double-glazed windows, clamps, gaskets, rubber

Table III. Glass–Metal Combinations

Expansion Range	30–40	40–60	60–80	80–100	100–120
Metals	Silicon	Tungsten Molybdenum Kovar 42 NiFe	Titanium Tantalum 46 NiFe	Dumet Platinum 52 NiFe No. 4 alloy	Cast iron 430 446 Cold-roll steel
Glasses	KG–33 7740 EE–5 7200	EN–1 7052 EN–4 7056 EN–5 7040 ES–1 7050 EE–2 1720 K–650 3320 K–772 7720	N–51–A 5630 N–10 7280 IN–3 7520 RP–3 7530	Kg–1 0120 KG–12 0010 R–6 0121 TM–5 1781 TM–9 8160 Tl–2 9010 EG–19 9013	TL–21 TL–27 TL–28 TH–10 9010 9019 1780
Solder glasses		SG–7 7574 CV–635 1826	CV–111 XS–1175–Mi XS–1190 SG–95 SG–200 SG–202 CV–285	SG–67 7570 SG–100 7575 CV–135 7572 CV–455 CV–97 CV–101	CV–605 CV–130 CV–808HD SG–68 CV–9 7595 7572 8363 7595

Table IV. Metal Preparation

Cleaning	Solvents, degreasing, washing, rinsing
Sandblasting	Descaling, alumina grit
Degassing	Cracked ammonia, hydrogen
Oxidation	Flame, furnace-selected atmospheres
Chemical treatment	Borating, fluxing

Table V. Sealing Processes

Flame	Gas, hydrogen, gas + electrode
Furnace	Jigs, gas or electric
Lehr	Cer-DIP, frame
rf sputtering	Metal flanges, disks, caps
ir	Reed switches, iodine lamp

Metal Preparation

Preparation includes one or more of the processes shown in Table IV. Specific procedures were developed for each metal part of the glass–metal joint. The common processes are cleaning and sandblasting. Generally, trichloroethylene abrading by alumina grit, sandblasting, and vapor degreasing are used for all metals. Degassing, oxidation, and pretreatments vary with different metals. Kovar* is frequently degassed in hydrogen or cracked ammonia for $\approx$15 min at 1150°C (2100°F). Kovar oxidation is by direct flame, rf sputtering, or in a furnace at 930°C (1700°F) for 4 min in air. The No. 4 and stainless steel alloys are oxidized in wet hydrogen at 1315°C (2400°F) for 20 min, followed by air and flame heating. Copper preparation is the most complicated procedure, including hydrogen firing and acid etching with several rinsings. Copper oxidation includes borating to chemically formed cuprous oxide. Molybdenum, tungsten, and tantalum are pre-oxidized by direct-flame preheating. The last step of metal preparation is chemical treatment. Some frequently used treatments include fluxing; borating; bright-dip; and copper, nickel, tin, silver, gold, and cadmium plating.

Processing

The actual glass–metal sealing process is determined by the final specification of the desired product. Table V shows processes used industrially.

Annealing

Glass–metal seals are generally annealed at $\approx$10°C above the glass annealing point for $\approx$15 min. Normally, hot seals are placed into preheated furnaces. For Dumet seals, the fast-air cooling cycle (tempering) is preferred.

Inspection

Annealed seals are inspected for specific characteristics required by the customers' specifications. Many seals are checked for vacuum and hermeticity by helium-leak detectors. Electrical-resistivity, current-leakage, and high-voltage-breakdown tests are also used frequently for inspecting electronic parts. Feed-throughs are tested by pull for mechanical strength. Other required tests include those for thermal shock, pressure, vibration and corrosion resistance, solderability,

---

*Westinghouse Electric Co., E. Pittsburgh, PA.

and oxide stripping. Glass–metal seals must pass these tests to serve well in their intended function.

Solder Glasses

Solder glasses are low-melting glasses suitable for sealing assemblies, but they may deform at high temperatures (e.g., in color televisions), or the electronic device may be damaged by heat exposure (in phosphors, silicon, or MOS devices). They can be logically grouped as *vitreous* or *crystallizing,* and both groups can be *single-* or *multicomponent.*

Vitreous solder glasses remain glassy during and after sealing, but the crystallizing glasses change into ceramic structures with crystals and a glassy phase. Furthermore, these two solder glasses differ in desired properties, making them applicable for various sealing processes and for final products with specific seal specifications.

Vitreous solder-glass seals have several advantages.

1. Salvaging of parts is achieved by remelting the sealed joint without significantly changing the glass properties.

2. Annealing can adjust the seal joint stresses.

3. Expansion and contraction can be adjusted by additives.

4. Sealing time decreases with increased peak holding temperature.

Two disadvantages of the vitreous seals are that they are slightly sensitive to thermal shock, due to cooling temporary stresses, and that chemical corrosion is usually somewhat higher than for the crystallizing sealing glasses. Neither of these disadvantages seriously limits their applications.

Without available crystallizing solder glasses, color television bulbs could not be exhausted and evacuated at desired temperatures. The main advantage of these glasses is their rigidity at temperatures not far from the sealing temperature. Chemical corrosion resistance and mechanical shock resistance are usually higher than for vitreous solder seals. Although the crystallizing solder glasses accomplish their intended mission very well, only well-controlled processing achieves desired crystalline structures.

Application of Solder Glasses

Presently, solder glasses are used by the color-television and the electronic hermetic-packaging industries.

Color-Television Applications

Selected solder glasses, e.g., Table VI, are sold in powder form, which the customer processes into a relatively thick paste by adding selected vehicles. This paste, with a high glass content, is extruded by dispensers on the funnel part of the assembly. The deposition process is controlled for high production with miniscule losses (1/100). The controlled shape, spacing on the sealing edge, and paste weight are achieved daily. These weights may be within ±2 g in control. Using dried solder-glass paste, a funnel is placed into a proper fixture with the neck (narrow portion) down. The face is then placed in contact with the solder glass in a desired orientation maintained by fixtures, and this assembly is sealed in a lehr. The following is a typical schedule.

Heating rate ranges from 7° to 10°C per minute, minimizing the thermal shock of relatively thick glass parts with somewhat higher expansion and ensuring complete burning-out of organic vehicles. Incomplete removal of vehicles could chemically reduce the lead glasses and they, in turn, could cause high-voltage punctures.

Table VI. TV Solder Glasses

	CV–808HD	1307B	1304	8363
Density, g/cm^3	6.56	6.3	6.3	6.42
Expansion, $\times 10^{-7}$/°C	100	99	99	100
Contraction, $\times 10^{-7}$/°C	106	102	102	104
Annealing point, °C	300	330	320	304
Softening point, °C	370	400	400	375
Volume resistivity				
at 250°C, log $\Omega \cdot$cm	8.1	7.8	7.7	9.2
at 350°C, log $\Omega \cdot$cm	6.6	6.4	6.4	7.5

At ≈500°C the solder glass fluxes (i.e., reacts with, or dissolves) both glass parts of the color-television assembly. After about 10 minutes at peak temperature, the original vitreous glass begins to crystallize, and the fluxing action slows down. Crystallization takes about 20 to 28 min to achieve the desired strong structures. Then the assembly is cooled at 5° to 10°C per minute, eliminating any breakage from thermal shock.

Hermetic Packaging Cer-DIP

Hermetic packaging includes two ceramic parts (base and lid), a metal-lead frame, and two solder-glass sections which make the final seal. Examples are shown in Tables VII and VIII.

The supplier to this industry preglazes the two ceramic parts by silk-screen printing and then glazing the ceramic parts in lehrs in an air atmosphere, to remove organic vehicles. Temperatures range from 320° to 430°C and times from 4 to 12 minutes.

The lead frame of the proper metal (Kovar–42, 45 alloys) is sealed to the ceramic base by rapidly heating the glazed ceramic base above the recommended final sealing temperature (30°–50°C for 2 min). The lead frame is pushed into the molten glass and this subassembly is cooled. Next, the silicon-active device is bonded into the ceramic cavity, and the thin wires are attached to it and the lead frame. The "closing" sealing part of the process seals the active device by attaching the lid to the base. Heating rate varies from 20° to 140°C per minute, and peak temperatures from 390° to 500°C, with hold times from 4 to 12 minutes. The last step of the process, cooling, ranges from 7° to 60°C per minute.

The lead frame is tin plated and the DIP assembly, including the sealing glass, immersed into acid descaling solutions. Solder glasses with relatively high lead-oxide content are normally subjected to sulfuric acid. Hydrochloric acid is used for glasses with relatively low lead-oxide content (e.g., SG–200 and SG–202). Nitric acid is used only in very low concentrations (e.g., 2%), for minimizing any solder-glass removal.

Table VII. Cer-DIP Solder Glasses

	CV-111	XS-1175-Mi	XS-1190	SG-95	SG-202	SG-200	LS-0802	LS-0803	LS-0120
Density, g/cm³	5.85	4.25	4.75	6.8	5.2	5.6	6.78	7.19	6.92
Thermal expansion, 30°–250°C Contraction, ×10⁻⁷/°C	70	74	74	73	70	63	77	67.5	67.5
Annealing point, °C	318	305	305	310	307	295	310	300	315
Softening point, °C	380	350	350	370	355	345	360	350	385
Volume resistivity									
at 150°C, log Ω·cm	9.7	11.5	11.5	12.0	11.7	11.1	10.5	11.0	11.5
at 250°C, log Ω·cm	8.6	9.1	9.1	9.5	9.2	8.8	8.5	8.9	9.0
Dielectric constant									
at 25°C, 1 MHz	16	12	12	40	12	16	31	35	31
Alpha radiation, counts/cm²/h		2	0.2	10	1.0	0.2			

Table VIII. Special Solder Glasses

	SG-7	SG-67	CV-432	SG-100	CV-455
Density, g/cm^3	4.07	5.38	6.55	6.7	5.9
Expansion, $\times 10^{-7}/°C$	41	83	117	81	86
Contraction, $\times 10^{-7}/°C$	60	102	127	90	89
Annealing point, °C	469	365	290	312	315
Softening point, °C	571	441	327	361	365
Volume resistivity					
at 250°C, log Ω·cm	12.4	11.1	6.6	11.4*	10.4*
at 350°C, log Ω·cm	10.5	8.9	5.0	9.0	8.3
Dielectric constant					
at 25°C, 1 MHz	8.2	12.5	27.3	31.8	18.3
Applications	Kovar EN-1 N-5I-A S.S. alloys	Soda lime Platinum #4 alloy	Metals Irons Nickel Aluminum Coppers	Glass Panels	Glass Panels
	Low exp.	Med exp.	High exp.	Vitreous Med exp.	Crystallizing Med. exp.

*At 150°C.

Glasses in Microelectronics in the Information-Processing Industry

R. R. Tummala and R. R. Shaw

IBM Corp.
East Fishkill, NY   12533

Some specific applications of glasses in information processing, transfer, storage, display, and printing are described. General requirements and challenges for future glass application are discussed.

Worldwide, information processing is a huge industry. Innovations and new applications of technology are trademarks of this industry, as competition provides the spur to constantly improve and extend the performance of systems and components. Glasses have played critical roles in information processing at various stages, and they are expected to continue contributing to future developments. This review describes some previous glass applications and discusses the general requirements and challenges of future applications. For brevity, specific examples are limited to International Business Machines applications. Five major areas of information-processing technology are covered:

1. Information processing.
2. Information transfer.
3. Information storage.
4. Information display.
5. Information printing.

Information Processing

Information processing is popularly known as *computer crunching*. Logic and memory functions are carried out by microcircuits built on chips, which may be silicon, gallium arsenide, or other similar materials. Other alternatives such as Josephson junctions, operating at cryogenic temperatures, are also feasible. All of these components are referred to as "devices," to distinguish them from the "packaging" which surrounds them. Glasses are very important for both device and packaging applications.

Devices

Glasses are used primarily as electrical insulators, passivation layers protecting the device against hostile environments, and key process step components (diffusion masks, alkali getters, solder dams, crucibles for crystal growth, topography smoothers, substrates for photomasks, etc.). They are used most commonly as thin films or layers, and consequently relatively small amounts are required. Glass quality, however, is all-important: The purity and reproducibility are critical. Primary requirements of the finished layers are good adhesion, thermal compatibility, and low defect density. The layers must not contaminate the chip

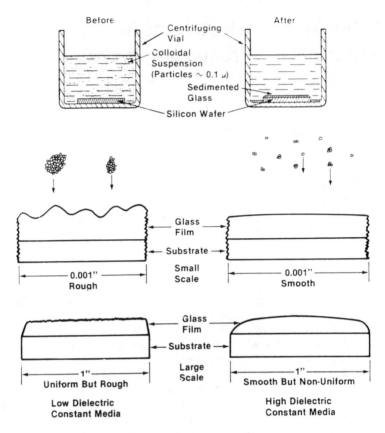

Fig. 1. Schematic of sedimented glass on silicon.

junctions (alkali ions are particularly bad), must offer high migration resistance, and must be deposited at relatively low temperatures. Processes that have worked well for applying glass layers include sedimentation by centrifuging, rf sputtering, electron-beam evaporation, and chemical vapor deposition.

Figure 1 describes the sedimentation process.[1,2] Submicrometer glass particles in a colloidal suspension are spun down on a silicon wafer by the high artificial gravity (more than 1000 *g*) of a high-speed centrifuge. When the suspension medium has a low dielectric constant, there is a tendency toward agglomeration of the depositing particles, and a uniform but locally rough surface is obtained. If the dielectric constant of the medium is high, the particles are dispersed better, but the layer is not as uniform. A two-layer system of high- and low-dielectric-constant media achieves very high-quality depositions.[1] In other applications, a single medium (10% isopropyl alcohol/90% ethyl acetate) with a medium dielectric constant can achieve good results. The technique is useful for depositing $\approx$1-μm films of borosilicate and lead-borosilicate glasses. Consolidation and fusion to a coherent layer are achieved by short holds ($\approx$5 min) near the softening temperature.

Although rf sputtering has successfully deposited both borosilicates and silica,[3-5] dc sputtering is unsuccessful because of the poor electrical conductivity of the glasses. The rf-sputtered films closely resemble thermal oxide, or SiO_2 grown by thermally oxidizing Si in situ.[6] Typically, the films are slightly oxygen-deficient and may contain entrapped sputtering gas,[6] but there is no apparent effect on the performance of the film.

Electron-beam evaporation can deposit glass layers up to 50 μm thick.[7] Difficulties with this approach include compositional variations, low density, oxygen deficiency, and high residual stress. Most of these problems, however, can be overcome with careful attention to glass composition, post-deposition annealing, temperature control during deposition, and multisource evaporations.[7–9] This approach offers the benefit of high-rate deposition, approaching 1 μm/min.

Chemical vapor deposition (CVD) of silica, borosilicate, borophosphosilicate, and phosphosilicate glasses can be done between 250° and 1100°C, and 400°C is a fairly typical level.[3,6,10–12] Thicknesses to $\approx$10 μm are obtained. Annealing in a humid environment is beneficial for relieving strain and increasing density.

Other layer-glass-forming techniques of potential interest are the "chemical" routes, which include sol-gel, alkoxide hydrolysis, organometallic reactions, and so forth. Many industrial, government, and university laboratories are investigating this approach.

Packaging

Packaging of devices has become more than a basic necessity to support and protect the chip and offer it access to the outside world. In addition to these requirements, it is now necessary, in many cases, to take into account the effect of the package on system performance. That is, the package is no longer just an inert component, but a dynamic, interactive participant in signal processing, as is particularly clear in large-scale computer technology.[13] The packaging-materials set has a direct impact on the electrical environment of the wiring network and affects the time-of-flight of the signal pulse, on which the operating performance of the processer ultimately depends.

Glasses are used in packaging as densifying aids, seals, and multilayer dielectrics. Densifying aids were first used in conducting pastes, which are screened onto substrates to form signal- and power-connection wiring. The earliest applications were in Ag–Pd pastes, which used a lead-borosilicate glass of composition (in wt%) $30SiO_2$-$50PbO$-$10B_2O_35Al_2O_3$-$2Na_2O$ and three other oxides. This was screened onto an alumina substrate and sintered at 750°C. The glass helped control the shrinkage behavior of the pastes during drying and firing, and aided in bonding the metals to the substrate. Later, other glasses were used for the same purpose, with W- and Mo-containing pastes. In substrate applications, high-alumina substrates were densified below 1600°C[14,15] using a 10% glass phase. This application required extremely close control of the firing shrinkage; satisfactory control was achieved by careful attention to the composition of the glass phase, which was a silicate glass containing CaO and MgO.

Figure 2 shows the effect of the total CaO and MgO content on the overall shrinkage behavior of the laminated substrate. Very close control is possible only with close tailoring of the glass compositon.

Seal glasses have wide application in packaging. Seals may be required to hermetically encapsulate a component, nonhermetically bond components together, or provide mechanical connections and support structures. All combinations of glass, metals, and ceramics may be involved in a given design, and the appropriate glass-seal composition must be carefully considered in view of the differing thermal properties of the various materials. For matched seals, the sealing process

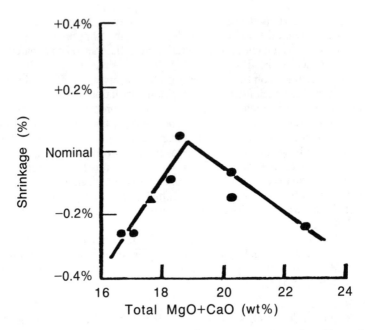

Fig. 2. Variation of substrate firing shrinkage with composition of glass phase.

is relatively straightforward where there is a close match (less than a few hundred ppm difference in thermal expansion). Complications arise when larger differentials occur: Usual approaches are to design a compressive stress into the glass by buttressing it with a material of higher thermal expansion, using a series of glasses with different properties to form a graded seal, or designing in a relatively thin layer of a deformable metal that can relieve stresses in the glass.

Glasses[16] and glass-ceramic mixtures[12-15,17,18] also can be used successfully as dielectric layers in complex multilayer ceramics (MLCs). Power and signal wiring is incorporated within the MLC, providing a combination chip carrier and space transformer and thus reliable connections between the chip and the outside world. The MLC simultaneously provides a favorable electrical environment and allows the package to be adequately cooled. Up to 150 layers have been reported.[18] Figure 3 shows various wiring patterns screened onto individual layers, and Fig. 4 shows a large MLC populated with chips.

Information Transfer

Glass optical fibers are now an extremely important means of transferring information. Acting as waveguides, they offer the advantages of extremely high bandwidth (a measure of transfer-rate capability), high resistance to electromagnetic interference, and good security against being tapped. Optical-fiber installations for communications, data busing, teletype, video, closed-circuit television, etc., now exist worldwide and are increasing rapidly. Three trends have had a major impact on the acceptance of glass fibers for these purposes.

1. Transmission loss of the glass has dramatically decreased: Twenty years ago, a loss figure of 10–100 dB/km could be achieved only at great effort on a

Fig. 3. Various screened greensheet patterns.

Fig. 4. Substrate with attached chips.

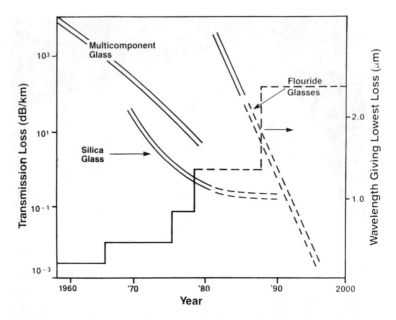

Fig. 5. Trends of optical-fiber transmission loss and preferred wavelength.

laboratory scale over selected lengths. Today's laboratory figure is 0.154 dB/km at 1.55 μm,[19] with commercially available fiber in the 2- to 4-dB/km range.

2. The capabilities of the launcher and receiver components of an optical transmission system have similarly improved: Introducing solid state devices, such as light-emitting diodes and lasers for use as sources, and complex photodiode systems as receivers,[20] has greatly increased processing speed.

3. Due to the growing ability to understand and control loss mechanisms in the glass, coupled with source/detector improvements, the transmission region has been shifted to longer wavelengths: This progression favors lower losses, because of lower scattering at longer wavelengths. Figure 5 shows these trends of lower loss and longer wavelength vs time. At present, single-mode silica waveguides have substantially lower loss than do multimode, multicomponent glasses. However, systems using fluorine-containing glasses may be better than either in the future. The preferred wavelength is presently $\approx$1.3 μm and expected to shift out to the 2.5-μm region in the future.

Information Storage

Data have been stored in either a temporary or archival mode in a wide variety of technologies. Figure 6 indicates the trends of commonly available storage devices, in terms of access time vs storage capacity. A more desirable trend would follow the direction of the arrow.

Magnetic disk storage has been very popular for many years. Here, a ferrite recording head is suspended on an air cushion just above the recording disk. The general construction of the head is shown in Fig. 7. Two types of glass are required to construct the head: The first is a high-temperature composition for filling the gap in the nickel-zinc ferrite used as the active magnetic component; the second is a

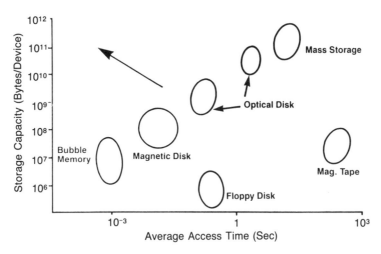

Fig. 6. Technology trends in various storage media.

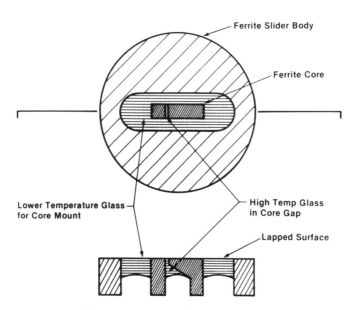

Fig. 7. Structure of ferrite recording head.

lower-temperature glass for bonding the head to the slider. The gap glass must match the thermal expansion of the ferrite and have a working temperature of $\approx 900°C$ and a very steep viscosity–temperature characteristic, so that the bonding glass cannot react extensively with it. The bonding glass also must thermally match the ferrite (expansion coefficient $= 90 \times 10^{-6}/°C$), wet both the ferrite and ceramic, and bond below the annealing point of the gap glass. These are difficult simultaneous requirements, but the challenge was met by the compositions shown

93

Table I. Recording-Head Glasses

Component	High-Temp. Gap Glass	Bonding Glass
	wt%	
BaO	42	0
B_2O_3	10	6
SiO_2	35	15
Al_2O_3	5	0
CaO	8	0
PbO	0	73
Thermal expansion coefficient, $\times 10^7/°C$ (*RT*–300°C)	80	80
Softening point, °C	770	490
Bonding/gap temperature, °C	900	600

Table II. Cathode Ray Tube Glasses

Component	Faceplate	Typical Seal
	wt%	
SiO_2	64.0	1.0
Al_2O_3	3.5	2.0
Na_2O	7.5	0
K_2O	10.5	0
BaO	8.0	0
PbO	0	77.0
CaO	3.5	0
MgO	1.5	0
B_2O_3	0	9.5
ZnO	0	10.5
Thermal expansion coefficient, $\times 10^7/°C$ (*RT*–300°C)	98.5	98.0
Softening point, °C	680	372
Annealing point, °C	490	

in Table I. A barium-borosilicate glass filled the gap-glass requirements, and a relatively simple lead-borosilicate composition served well for the bonding glass.

Information Display

The most popular information-display device is the cathode ray tube (CRT). This tube is almost entirely a glass-envelope device which is pressed or blown to

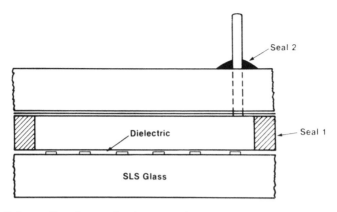

Fig. 8. Schematic of gas display panel.

shape and usually depends on sealant glasses. Typical faceplate and seal-glass compositions are shown in Table II.

An alternate large-area display device is the gas-discharge display panel.[21] The gas panel display is flat and takes up relatively little space; the panel excites gas cells by means of high-voltage electrodes distributed throughout the face. Gap spacings and seal requirements are very stringent.[21,22] Figure 8 shows a schematic of a gas panel and indicates the dielectric and sealing glasses. The substrate glass itself is a soda-lima-silica (SLS) glass, with a softening temperature of 725°C. The dielectric and seal glasses must have the same thermal expansion as the substrate, but the softening temperature of the dielectric glass must be intermediate between the substrate and sealing glasses. In addition, the dielectric glass must be able to form defect-free layers only 25-μm thick, over areas >300 × 400 mm, and at temperatures not exceeding the 600°C flow temperature of the SLS glass. The seal glasses must seal below the glass-transition temperature of the dielectric glass. The glasses developed to meet all of these simultaneous requirements were lead and lead-alkaline borosilicates; compositions are listed in Table III. Glass 1 was the successful candidate for dielectric application. Glass 2 represents a commercially available glass that came closest to the requirements, except for a too-low flow temperature. Glass 3 was successfully developed as a seal glass in cane form, replacing the commercially available glass 4, which was not available in that form. Glass 5 was a later development, incorporating crystalline beta eucryptite ($Li_2O \cdot Al_2O_3 \cdot SiO_2$) for a lower thermal expansion coefficient.[23]

Information Printing

Displaying the desired information is often not enough; it must also be printed or hard-copied for further work. Very high-speed printers have been developed for printouts, with rates of 20 000 lines per minute, corresponding to several hundred pages per minute.[24] Such rates are far too high for contact printing, and ink-jet printing is often resorted to; this approach can yield up to 1000 pages per minute.[25] Figure 9 shows a schematic of the ink-jet printing process, in which a high-frequency piezoelectric crystal breaks up a stream of ink into tiny droplets. These droplets are electrostatically charged and passed between deflection plates which electronically form the desired character on the paper. The character data input is part of the charging step. If no characters are to be formed, the droplets are left

Table III. Dielectric Glasses

Component	Glass 1	Glass 2	Glass 3*	Glass 4[†]	Glass 5[‡]
			wt%		
PbO	56.0	73.0	66.0	77.0	73.1
B_2O_3	21.5	12.5	14.0	9.4	9.5
SiO_2	12.0	14.3	2.0	1.0	1.9
Al_2O_3	1.0	0.2	3.5	2.1	0.5
CaO	5.5	0	0	0	0
MgO	2.0	0	0	0	0
Na_2O	2.0	0	0	0	0
Bi_2O_3	0	0	1.5	0	0
ZnO	0	0	10.5	10.5	0
CuO	0	0	2.5	0	0
beta eucryptite	0	0	0	0	15.0
Thermal expansion coefficient, $\times 10^7/°C$ (*RT*–300°C)	86.5	84.0	84.0	89.0	77.0
Softening point, °C	542	473	413	370	350
Seal temperature, °C	480	435	470	450	425

*Vitreous cane seal glass.
[†]Devitrifying seal glass.
[‡]Composite glass.

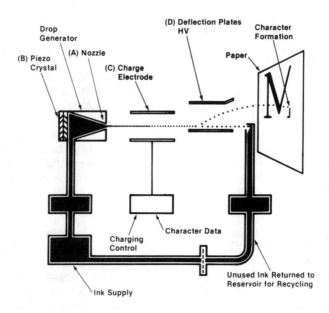

Fig. 9.   Schematic of ink-jet printer.

Table IV. Glass-Nozzle-Industry Standard

Component	Wt%
SiO_2	71.0
ZrO_2	15.0
Na_2O	11.0
K_2O	1.0
Al_2O_3	2.0
Thermal expansion coefficient, $\times 10^7/°C$ (RT–300°C)	64
Softening point, °C	869
Annealing point, °C	627
Weight loss, mg/cm^2, in electrostatic inks (24-h accelerated test at 70°C)	0.03

Table V. Comparison of Potential Nozzle Glasses Tested in Alkaline Electrostatic Inks

Component	Soda-Lime-Silica Glass	Borosilicate Glass	Improved Glass
	wt%		
SiO_2	73.6	81.0	58.5
ZrO_2	0	0	19.0
Al_2O_3	1.0	2.0	0
B_2O_3	0	13.0	0
CaO	5.2	0	0
MgO	3.6	0	4.0
Na_2O	16.0	4.0	16.0
K_2O	0.6	0	2.0
As_2O_3	0	0	0.5
Thermal expansion coefficient, $\times 10^7/°C$ (RT–300°C)	93	33	88.1
Softening point, °C	700	820	883
Annealing point, °C	525	565	673
Weight loss, mg/cm^2 (24-h accelerated test at 70°C)	0.324	0.288	0.024

uncharged, and fly through the deflection plates to a recycling reservoir. An array of such ink jets makes up the high-speed printer.

The glass nozzle is a critical part of the printer. It must allow precision fabrication through glass drawing, maintain stable dimensions by resisting the corrosion of the ink, and be sealable to form a nozzle array. An alkali-zirconium-silicate glass has been the industry standard; composition and properties are shown in Table IV. Improvements were required, however, and alternate glasses were developed. The composition and properties of some of the alternates are shown in Table V for alkali resistance and in Table VI for acid resistance. High corrosion resistance to both types of ink was desired. Two glasses, a borosilicate and an improved high-zirconium silicate, were found to have significantly improved corrosion resistance in short-term tests. However, as shown in Fig. 10, corrosion rates varied with time, and only the "improved" glass showed satisfactory performance in acidic media.

Summary

Glasses play key roles in many electronic and information-processing technologies, as indicated by the selected examples; many other instances could also be cited. In general, the glasses chosen for these applications must satisfy more unique requirements than do conventional glasses. First, such glasses are never used alone, but instead are combined with other materials, forming a functional system or subsystem. Second, relatively small quantities are required for most

Table VI.   Comparison of Potential Nozzle Glasses Tested in Acidic Electrostatic Inks

Component	Soda-Lime-Silica Glass	Borosilicate Glass	Improved Glass
	wt%		
SiO_2	73.6	81.0	58.5
ZrO_2	0	0	19.0
Al_2O_3	1.0	2.0	0
B_2O_3	0	13.0	0
BaO	0	0	0
CaO	5.2	0	0
MgO	3.6	0	4.0
Na_2O	16.0	4.0	16.0
K_2O	0.6	0	2.0
As_2O_3	0	0	0.5
Thermal expansion coefficient, $\times 10^7/°C$ (RT–300°C)	93	33	88.1
Softening point, °C	700	820	883
Annealing point, °C	525	565	673
Weight loss, mg/cm^2 (24-h accelerated test at 70°C)	0.0105	0.0046	0.0053

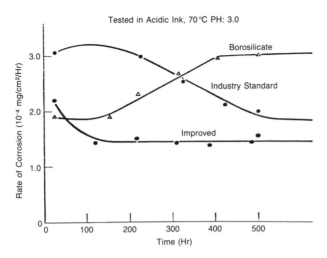

Fig. 10. Corrosion behavior of glass ink-jet nozzles.

applications. Quality demands for purity, reproducibility, and zero defect levels are, however, extraordinary. Third, almost all applications involve careful consideration of the bonding stress in the glass components. The short- and long-term bonding stresses give rise to the greatest single critical concern: inadequate long-term mechanical integrity of the product. This concern is itself not unique to the glass applications considered in this review, but is, of course, widespread in all glass applications. The consequences of mechanical failure are, however, extremely severe for information processing. Field failures of devices, for example, can immobilize a complex computer installation, which in turn might severely affect the operations of the user. Loss of an optical transmission line by mechanical failure could result in very expensive, time-consuming repairs. It is to the credit of designers and manufacturers that such problems are extremely rare: CRT glass envelopes do not implode, glass waveguides do well in hostile environments, MLC packages have established excellent reliability records, etc. These records are good, but much improvement remains possible. A much better understanding of the strength behavior of glasses in various applications is needed.

Challenges in Glass Science

Strength problems are important, but many other areas in glass science and technology demand better understanding also. Glasses for future information-processing applications should benefit from advanced understanding in the following areas:

1. *Particulate Technology:* The ability to generate, separate into close fractional sizes, and control the behavior of particles is increasingly important. The "green" body often must undergo extensive processing before final firing, and must possess many of the same qualities (low defect level, reproducibility, stability) as the final glass product. Understanding small-particle behavior in both the dry and suspended states is necessary for achieving these qualities.

2. *Organic-Inorganic Reactions:* Organic media are commonly used as carriers, deflocculants, binders, plasticizers, etc., in green-glass bodies. In many applications, it is necessary to remove virtually all organics from the final product.

99

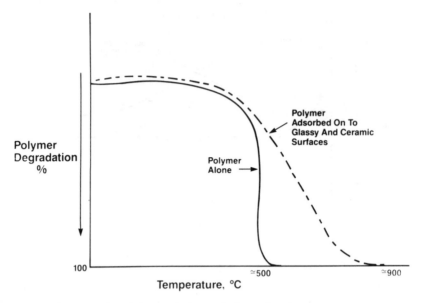

Fig. 11.   Schematic of thermal degradation of polymers.

The breakdown and removal of organics during firing is often complicated, however, by interactions between organics and glasses, as shown schematically in Fig. 11. Here, the degradation rate of the polymer phase is greatly affected by the presence of glass and ceramic surfaces, and the organic can be quite difficult to remove by thermal means; sintering behavior and final properties of the glass can be strongly affected by this problem. More work is needed on establishing the nature of the interaction and how it affects the degradation pathway.

3. *Sintering and Microstructure Control:* The quality of the green body strongly affects sintering behavior and microstructure. Since the resultant quality may be dominated by the green microstructure, considerable work has gone into pre-firing technology. One interesting approach is to work with nearly monosized spherical particles for better control of the green microstructure and sintering behavior.[26] Microstructure control is one of the most important areas requiring advanced work.

4. *Glass-Metal Bonding:* This bond, the most common in microelectronic applications, represents one of the most serious stress concerns. Future applications will depend heavily on the successful understanding, modeling, and design of this key interface.

5. *Fracture and Wear:* Fracture and wear behavior under various loading rates and atmospheres has been investigated for many common glasses, but generally not for the compositions used in microelectronics. Since numerous failure mechanisms are possible, work is needed to establish which mechanism is dominant and to devise ways of improving product performance. One interesting approach is to incorporate ZrO_2 particles within glasses, increasing the fracture stress through phase transformations in the vicinity of the crack stress field. Other inclusions have also been useful, as have chemical and thermal tempering. The overall goal in fracture improvements is to use a compressive stress for counter-

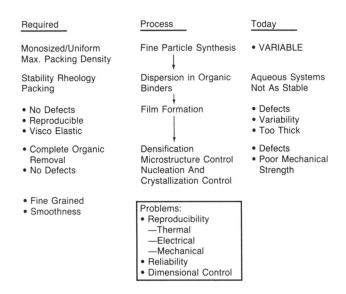

Required	Process	Today
Monosized/Uniform Max. Packing Density	Fine Particle Synthesis	• VARIABLE
Stability Rheology Packing	Dispersion in Organic Binders	Aqueous Systems Not As Stable
• No Defects • Reproducible • Visco Elastic	Film Formation	• Defects • Variability • Too Thick
• Complete Organic Removal • No Defects	Densification Microstructure Control Nucleation And Crystallization Control	• Defects • Poor Mechanical Strength
• Fine Grained • Smoothness	Problems: • Reproducibility —Thermal —Electrical —Mechanical • Reliability • Dimensional Control	

Fig. 12.   Typical electronic glass and glass-ceramic dielectric films.

acting the applied tensile stresses to which glasses are sensitive. Wear behavior is related to strength and fracture considerations, but is also very sensitive to local temperature, pressure, and chemical conditions.

6. *Grain-Boundary Properties:* Glasses play several important roles in grain boundaries. In glass-bonded substrates, the overall sintering and shrinkage behavior has already been shown to depend heavily on glass composition. Thermal stresses must be accommodated during temperature cycling. The overall properties of the final product (electrical, physical, magnetic) are affected by boundaries. As the particle size decreases, which is the trend for many electronic applications, the relative importance of understanding grain boundaries will increase.

7. *Thin-Film Insulators:* Glasses are well suited for making thin- and thick-film structures in microelectronics. Problems still exist in reproducibility, reliability, and dimensional controls, and represent one of the more demanding areas. As an example, Fig. 12 describes present difficulties in producing a high-quality dielectric film using particulate technology. It is evident that many deficiencies exist in current practice, ranging from particle production through dispersion, film formation, densification, and crystallization. These deficiencies are fairly representative of many thin-layer insulator structures, which are very important in microelectronic applications.

Conclusions

There are many important applications for glasses in the microelectronics and information-processing industries. Many more demands will undoubtedly be made on glasses as new products and technologies are developed. It is interesting to compare this brief review with a somewhat similar overview made almost 20 years ago.[27] The trends and types of glass applications at that time are familiar: Relatively small amounts of high-quality material were required, a trend toward miniaturization was evident, and there was a need for developing new compositions to solve the problems of new technologies. Problems reported at that time included dimensional stability, surface quality of glass substrates, and stability of optical components during operation — again, a familiar list. With that perspective, there is little doubt that glasses will be called on frequently for high-technology applications in the future, and that there will be ever-increasing demands for greater quality and reproducibility.

References

[1] W. A. Pliskin and E. E. Conrad, "Techniques for Obtaining Uniform Thin Glass Films on Substrates," *Electrochem. Technol.,* **2** [7–8] 196–200 (1964).

[2] J. A. Perri, "Glass Encapsulation," *Solid State Technol.,* **5**, 19–23 (1965).

[3] G. L. Schnable, W. Kern, and R. B. Comizzoli, "Passivation Coatings on Silicon Devices," *J. Electrochem. Soc.,* **122** [8] 1092–1103 (1975).

[4] S. Singh, W. Bowner, I. Camlibel, W. Grodkiowicz, G. Pasteur, L. Vanuitart, and R. Williams, "Borosilicate Glass Films for InP Encapsulation," *Appl. Phys. Lett.,* **38** [5] 349–52 (1981).

[5] G. J. Griffiths and P. J. Khan, "Investigation of RF-Sputtered Nd-Glass Films for Integrated Optics," *J. Vac. Sci. Technol.,* **16** [1] 20–24 (1979).

[6] W. A. Pliskin, "Comparison of Properties of Dielectric Films Deposited by Various Methods," *J. Vac. Sci. Technol.,* **14** [5] 1064–81 (1977).

[7] K. C. Park and T. J. Weitzman, "E-Beam Evaporated Glass and MgO Layers for Gas Panel Fabrication," *IBM J. Res. Develop.,* **22** [6] 607–12 (1978).

[8] W. Fedrowitz and W. A. Pliskin, "The Evaporation Capability of Various Glasses," *Thin Solid Films,* **72** [3] 485–86 (1980).

[9] D. M. Sanders, E. N. Farabaugh, and W. K. Haller, "Glassy Optical Coatings by Multi-source Evaporation"; pp. 346–405 in SPIE Proceedings, Vol. 346. Tech. Sympos. East, May 1982.

[10] W. Kern, "Chemical Vapor Deposition Systems for Glass Passivation of Integrated Circuits," *Solid State Technol.,* **12**, 25–33 (1975).

[11] B. Mattson, "CVD Films for Interlayer Dielectrics," *Solid State Technol.,* **1**, 60–64 (1980).

[12] W. Kern and R. K. Smeltzer, "Borophosphosilicate Glasses for Integrated Circuits," *Solid State Technol.,* **6**, 171–79 (1985).

[13] W. B. Grobman, "High-End Computer Packaging — VLSI Scaling and Materials Science," *J. Vac. Sci. Technol.,* **A3** [3] 725–31 (1985).

[14] A. J. Blodgett and D. R. Barbour, "Thermal Conduction Module: A High-Performance Multilayer Ceramic Package," *IBM J. Res. Develop.,* **26** [1] 30–36 (1982).

[15] A. J. Blodgett, "Microelectronic Packaging," *Sci. Am.,* **249** [1] 86–96 (1983).

[16] R. R. Tummala, "Borosilicate Glass Dielectric Composition for High-Performance Glass-Metal Package," U.S. Pat. No. 3 640 738, 1971.

[17] Y. Shimada, K. Utsumi, M. Suzuki, H. Takamizawa, M. Notta, and S. Yano, "Low Firing Temperature Multilayer Glass-Ceramic Substrate," 33rd Electronic Components Conference, May 1983.

[18] M. Terasawa, S. Minami, and J. Rubin, "A Comparison of Thin Film, Thick Film, and Co-Fired High Density Ceramic Multilayer with the Combined Technology: T and T HDCM (Thin Film and Thick Film High Density Ceramic Module)," *Int. J. Hybrid Microelectr.,* **6** [1] 607–15 (1983).

[19] "Sumitomo Reports Record Low Loss for Fiber," *Lasers Applic.,* **5** [5] 50 (1986).

[20] D. G. Thomas, "Optical Communications," *Res./Develop.,* **6**, 199–204 (1984).

[21] T. A. Sherk and R. A. Rita, "Plasma Displays: An Alternative to LCD's and CRT's," *Res./Develop.,* **11**, 152–55 (1984).

[22] C. Perry, "Spacer Technology for an AC Plasma Panel," *Electron. Prod.,* **26** [5] 109–11 (1983).

[23] R. A. Rita, "Gas Panel Seal Glass," U.S. Pat. No. 4 478 947, October 23, 1984.

[24] R. C. Miller, Jr., "Introduction to the IBM 3808 Printing Subsystem Models 3 and 8," *IBM J. Res. Develop.,* **28** [3] 252–56 (1984).

[25] R. H. Darling, C. -H. Lee, and L. Kuhn, "Multiple-Nozzle Ink Jet Printing Experiment," *IBM J. Res. Develop.,* **28** [3] 300–306 (1984).

[26] H. K. Bowen, "A Near Perfect Hybrid Substrate, Soon?" *Circuits Mfg.,* **6**, 44–48 (1983).

[27] H. Rawson, "Report on the Symposium on Glass in Electronics," *Glass Technol.,* **7** [4] 115–20 (1966).

Section II

Glass Composition

High-Silica Glass

PETER P. BIHUNIAK

General Electric Co.
Lighting Business Group
Richmond Heights, OH   44143

This overview presents a brief historical review of the commercial development of high-silica glasses, focusing primarily on vitreous silica. This relatively simple glass, essentially SiO_2 plus impurities, provides a chemically clean system for amorphous-state inquiry, including property tailoring via doping. Consequently, these glasses have evolved into a family whose property range is determined by impurities (due to raw materials or processing) or intentional dopants.

In general, the relative chemical inertness (viz., acid durability), high thermal refractoriness (viscosity), low thermal coefficient of expansion (high thermal shock resistance), and high optical transparency have been the bases for a variety of products during the commercial lifetime of high-silica glasses. Table I details some of the more important current product applications.

While the high thermal refractoriness defines many product and processing applications, this high viscosity, along with a short working range, a tendency to devitrify, and a high vapor pressure, makes it a very difficult glass to manufacture. Normal commercial melting and forming processes simply do not apply and, consequently, many unusual and novel approaches have been developed. These processes will be put into historical perspective by chronologically reviewing highlights of the period preceding high-silica-glass commercialization which took place around the turn of the twentieth century. Many technological advances have occurred since then, but this review will be somewhat arbitrarily limited to a discussion of synthetic technology; Vycor,* or reconstructed glass, technology; continuous processing of tubes or rods; and raw-material beneficiation. As this review is limited to bulk glass methods, there will be no discussion of sol/gel techniques, a rapidly emerging technology that addresses all aspects of high-silica-glass manufacture.

Precommercialization Period

Not until the development of oxygen-injected torches early in the nineteenth century did high-temperature research and, therefore, high-silica-glass investigation become possible. Prior to this time, there simply was not enough concentrated energy to fuse crystalline quartz.[1]

Marcet, in 1813, performed the first documented small-crystal fusion using an O_2-injected alcohol lamp (see Table II). Clark, in 1821, fused quartz crystals with an H_2/O_2 torch.[2] However, it was not until 1839, with the effort of Gaudin,[4] who used an H_2/O_2 flame to fuse crystalline quartz, that properties were measured: In working drops and threads, he noticed the tendency to volatilize and devitrify. Furnace fusions date to 1856, when Deville[5] fused a relatively small sample in a

---

*Corning Glass Works, Corning, NY.

Table I. Application of High-Silica Glasses

Industry	Function	Primary Property
Semiconductors		
Polycrystalline Si mfg.	*p.v.*	Inertness, refractoriness
Monocrystalline Si mfg.	*p.v.*	Inertness, refractoriness
Epitaxial, chemical deposition, diffusion, oxidation	*p.v., m*	Inertness, refractoriness
Photolithographic	*s*	Expansion, uv transparency, hardness
Lighting		
High-intensity-discharge (xenon, mercury, halogen)	*r.c.*	Inertness, refractoriness
Optical		
Fiber optics	*p.v., s*	ir transparency, refractoriness
Lenses, windows, mirrors		Expansion, transparency
Chemical		
Labware	*p.v.*	Inertness, expansion, refractoriness
Tubing	*p.v.*	Inertness, expansion, refractoriness
Miscellaneous		
GC columns	*s*	Inertness
Tubing, fiber, wool	*m*	Inertness, refractoriness

p.v. = processing vessel, *m* = thermal muffle/shield, *s* = substrate, *r.c.* = reaction chamber.

Table II. High-Silica-Glass Precommercial Era

Date/Observer	Heat Source	Event Significance
1813/Marcet	Alcohol/O_2 flame	First recorded small-crystalline fusion
1821/Clark	H_2/O_2 flame	Lampworking
1839/Gaudin	H_2/O_2 flame	Lampworking, noted devitrification and vaporization problems
1856/Deville	Coke furnace	Furnace fusion of $\approx$30 g
1869/Gautier	H_2/O_2 flame	Lampworking of capillaries, bulbs, exhibited at 1878 Paris Exposition
1887/Boys	H_2/O_2 flame	Fiber draw, property observations
1888/Parsons	Resistance furnace	Demonstration of resistance fusion

Table III. High-Silica-Glass Commercialization Era

Inventor	Process Technique	Advancement
	Transparent (Fused Quartz)	
Shenstone (1902)	Crystal fritting	Transparent fusion
Shenstone, Kent (1903)	Arc fusion	Powder injected into arc, boule process, no vessel
Heraeus (1908)	H_2/O_2 torch	Boule deposition, no vessel
	Opaque (Fused Silica)	
Thompson, Hutton (1904)	Carbon resistance	Low-T fusion, reducing $C\text{-}SiO_2$ reaction
Bottomley (1904)	Carbon resistance	Molding/redraw of plastic ingot via controlling CO sheath during fusion
Vöelker (1910)	Carbon resistance	"Potato" process

coke furnace. Gautier, again using an H_2/O_2 flame, fashioned tubes, bulbs, and a thermometer and provoked interest by exhibiting these items in the 1878 Paris Exposition.[6]

Boys, in 1887, flame-fused and worked crystal,[7] fashioning fibers which he used as suspension wires for galvanometers. Such demonstrations provided further impetus for developing a commercial process that began around 1890. About that time, Parsons, Thompson, and others demonstrated resistance-furnace fusions, providing another method for technological innovation.[8]

The Advent of Commercialization

Commercial development of vitreous silica occurred primarily between 1899 to 1910 (see Table III) and was centered largely in England, France, and Germany.[9] Prior to this time fusions were rather small and on the scale of laboratory tinkering. Developing a commercially feasible technology required significant invention and innovation to overcome inherent high-temperature fusion problems (power input and vessel reaction) and fabrication difficulties presented by the system's refractoriness and tendency to devitrify.

Basically, there were two categories of material at this time: *transparent* and *opaque*. It is likely that the semantic confusion persisting today dates to this era, when it became commercially acceptable to distinguish between transparent and opaque vitreous silica by referring to the transparent glass, made by the fusion of selected quartz crystal, as *fused quartz* and the opaque glass, made from sand, as *fused silica*.[10]

Manufacture of transparent glass required the use of clear, selected, quartz crystalline raw material. However, the quartz polymorphic inversion at 573°C made transparent fusions very difficult because of splintering, air encapsulation, and subsequent bubble formation and retention in the melt. One solution involved the formation of a vitreous film to effectively seal the splintered mass from atmospheric encapsulation. Herschkowitsch[11] accomplished this via heating at

500°C, then plunging into a furnace held at white heat and Heraeus[12] via iridium vessel fusions and subsequent lampworking. Shenstone,[13] however, solved the problem by converting a disadvantage into an advantage: By heating the quartz crystal to red heat and water-quenching, he produced a crystalline frit. The small, clear crystals could then be flame-fused into rods which could, in turn, serve as feedstock to form tubes by softening the bound rods around a platinum-core rod. Tubes could then be lampworked into crucibles.

In 1903 Shenstone, and then Kent, explored arc fusions of this powdered crystal.[14] They developed the forerunner of a boule furnace, which eliminated the need for a potentially reactive vessel. The vessel-less approach by which powder was injected into a flame using a boule was further advanced by W. C. Heraeus.[15]

For opaque-ware manufacture, carbon electrode resistance fusion was a desirable method for obtaining adequate power input. However, the formation of CO during fusion was a serious obstacle. Thompson and others tried to minimize this CO formation via low-temperature fusions,[16] but Bottomley's solution was more clever.[17] Instead of avoiding this reaction, he took advantage of it by using a carbon rod to heat a sand mass to fusion. The CO generated would blow the molten plastic mass into a mold as the ends were kept cool to trap the gas. The melt was vessel-less in that the surrounding excess sand served as the container. The sand was also thermally insulating, maintaining the plasticity of the melt and thus permitting subsequent fabrication via redraw or molding techniques.

In a process variation developed by Vöelker,[18] CO-gas generation was augmented by the use of additional organic matter. This became known as the "potato" process.

With the technological basics advanced by Shenstone, Bottomley, Heraeus, and others, commercialization was well underway by 1910. Primary product applications during that era included lab ware, chemical processing vessels, instruments, and one of the few vitreous silica consumer items, gas-lamp chimneys.

Technological Advances

Significant mechanical processes were invented during the initial commercialization period. These batch processes were often purity-limited. Following this period, a number of significant, often proprietary, advances took place, addressing the issues of purity and continuous fabrication. Among the significant advances were synthetic technology, Vycor (reconstructed glass) processing, continuous tube and rod manufacturing, and raw-material beneficiation.

Synthetic Vitreous Silica

In a patent filed in 1934, Heany disclosed a method of producing what today is referred to as a high-purity "soot" by pyrolyzing silanes.[19] Fabrication consisted of slip-casting aqueous pastes of silica powder, a forerunner of sol technology.

Synthetic technology as we know it today, however, originated with the work of J. F. Hyde. In an invention filed in 1934 (see Fig. 1) containing 12 claims, Hyde described the $SiCl_4$-flame hydrolysis process.[20] He used many examples, including variations of Si-compound delivery to the torch, formation of tube preforms with subsequent vitrification on or off the target mandrel, and "soot" collection followed by pressing and sintering into articles. He produced a number of articles, including tubes, lenses, and prisms. This technique addressed two major processing disadvantages: high fusion temperatures and purity limitations. Although boule growth typically employed high temperatures (up to 1700°C), viscous sintering of high-surface-area (energy) particles permitted glass article formation as low as 1200°C. Such techniques have been advanced to achieve Schlieren optical-quality blanks

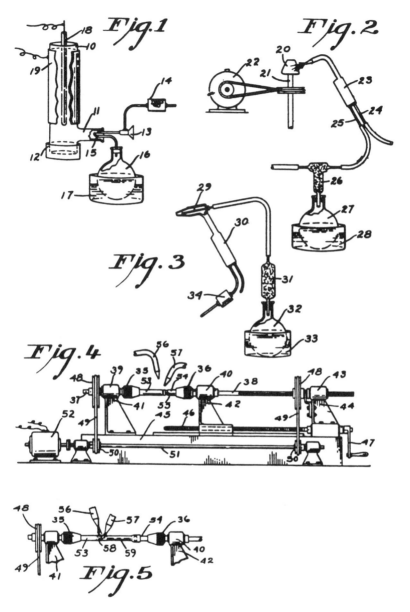

Fig. 1. Method for making a transparent article of silica (U.S. Pat. No. 2272342).

for precision mirror fabrication.[21] Furthermore, it was relatively straightforward to purify silicon-containing liquid compounds by fractional distillation, producing vitreous silica with improved uv transmission and resistance to radiation darkening. In addition, this chemical-fabrication process offered the possibility of property tailoring, by selective metallic-ion doping using volatile metallic compounds. Dalton and Nordberg studied a variety of dopants and advanced the TiO_2-SiO_2 family of zero coefficient of thermal-expansion glasses.[22] Such

109

techniques provide the basis for methods used today in manufacturing optical waveguides.

This flame hydrolysis approach was, however, a very "wet" process, because of the incorporated structural hydroxyl groups, OH^-, arising from the torch atmosphere. Typically, in situ glass fusions of this type contain approximately 1000 ppm OH^-, resulting in a strong ir absorption at 2.73 μm and a significant reduction in viscosity. To eliminate this contaminant, dry flames were investigated. The plasma-torch system, of current widespread commercial use, was first developed by Winterburn[23] and effectively reduces this "water" content to ≈ 5 ppm.

Reconstructed High-Silica Glasses

Another noteworthy advance by Hood and Nordberg was the development of the reconstructed-silica-glass process, the Vycor technique.[24] With clever microstructural manipulation, they eliminated the need for high fusion temperatures and the associated problems posed by such a refractory system. This novel approach consisted of first conventionally melting and fabricating a borosilicate composition. Following this procedure, the formed article would be heat-treated to $\approx 600°C$, separating interconnected high-SiO_2 (96–98%) and low-SiO_2 phases. The low-SiO_2 phase was then preferentially acid-leached, leaving behind the 96–98% SiO_2 skeleton. This so-called *thirsty glass* was then viscous-sintered to form a monolithic glass having physical properties close to those of vitreous silica.

Applications have been extensive and include lab ware, chemical-processing vessels, and high-intensity-discharge lamp arc chambers. Hood and Nordberg also demonstrated spectral-property tailoring by impregnating the thirsty glass with a metallic salt solution,[25] then drying and sintering. Such an approach has been updated recently, and was the basis of the Phasil process for optical-waveguide manufacture.

Continuous Drawing of Rods/Tubes

The high-pressure mercury-discharge lamp was successfully introduced in 1934, and with it came the high-volume need for a high-silica-glass arc chamber.[26] The process, first invented by Hänlein in 1939, can be viewed as a miniature, cool-crown, electric furnace, combining both fusion and forming in one tank[27]: In this process (see Fig. 2), a batch of crushed quartz crystal was continuously fed into the top of a resistance-heated refractory-metal (Ta, W, Mo) crucible, where it was fused and subsequently drawn through a refractory-metal die, over an inert-atmosphere-sheathed core, and into a tube.

Naturally, lampmakers were the primary developers of this continuous technique, which offers high-volume cost efficiency and good dimensional control compared to ingot-redraw techniques. Today, there are many variations of crucibles, nozzles, protective atmospheres, and power-input methods.

Property tailoring via batch doping can also be achieved in these furnaces. Today, in addition to lamp arc chambers, similar continuous techniques are used for manufacturing semiconductor processing rods and tubing and fiber-optic substrate tubing.

Beneficiated Raw Materials

For natural vitreous silicas, the availability of an inexpensive, impurity-consistent raw material has always been of major concern. Mined quartz crystal often has a variable chemistry. As the semiconductor industry grew, higher purity and improved refractoriness were required. Furthermore, there has always been some industrywide anxiety over supply availability. In 1974 this was underscored by the Brazilian government's decision to cut exports by two-thirds and increase

110

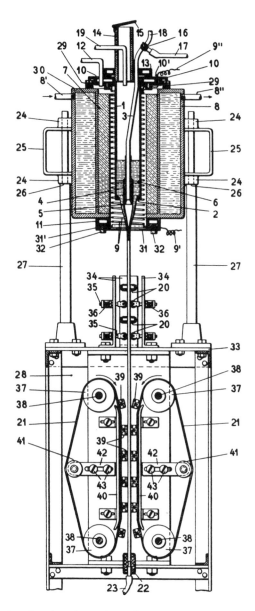

Fig. 2. Apparatus for drawing pipes from quartz or glass with high silicic acid content (U.S. Pat. No. 2 155 131).

prices by a factor of 10.[28] This led to the development of other sources, notably African.

Another approach taken by General Electric in the early 1960s was to innovate available mechanical, thermal, and chemical beneficiation techniques to render sand consistently high in purity and capable of transparent fusions, thus erasing the turn-of-the-century distinction between fused quartz and fused silica.[29] Today,

111

Table IV. Vitreous Silica: Commercial Categorization

		Raw Material, Increasing Purity			
		Sand (Alk ≈ 50, M ≈ 400, Al ≈ 200)	Crystal (Alk = 5, M < 10, Al = 10–50)	Beneficiated Sand (Alk = 5, M < 10, Al = 20)	Synthetic (Alk = 1, M ≈ 1, Al ≈ 1)
Flame Fusion: Decreasing OH⁻ →	CH₄, H₂/O₂		OH⁻ = 200 H/A—T08, Ultrasil, Homosil TSL/TAFQ—Vitreosil 055 TOS–T1030,1070,1130(d), 1170(d) Q/S—981		OH⁻ ≈ 1000 Cl⁻ ≈ 100 CGW–7940, U.L.E.(d) Dynasil GE(WQS)—Synsil H/A—Suprasil TSL/TAFQ—Spectrosil TOS–T–4040 Q/S—Tetrasil A, B
	Plasma				OH⁻ < 10 Cl⁻ < 200 H/A—Suprasil W TSL/TAFQ—Spectrosil WF TOS–T–4042 Q/S—Tetrasil SE
Elec. Fusion: Decreasing OH⁻ →	Arc atm			OH⁻ ≈ 30 GE–510 GTE crucibles QSI crucibles	
	Resistance/ induction atm	OH⁻ = 200 GE–318 H/A—Rotosil TSL/TAFQ—Vitreosil TOS–T100,200,800 Q/S—opaque		OH⁻ ≈ 30 GE–511 Pyro crucibles	
	Vac.		Q/S–453, Purposil, Pursil, Germiosil(d) 676	OH⁻ ≈ 30 GE–124,204	
			OH⁻ ≈ 30 TOS–T–2030		
	Rebake			OH⁻ < 5 GE–214,982,219(d) GTE–SG 255C TOS–T–7082	

Alk = Alkali content, ppmw total; M = transition metal content, ppmw; Al = aluminum metal content, ppmw; GTE = Sylvania; QSI = Quartz Scientific Inc.; Q/S = Quartz et Silice; H/A = Heraeus, Amersil; TSL = Thermal Syndicate Ltd.; TAFQ = Thermal American Fused Quartz; TOS = Toshiba; GE = General Electric; WQS (West Deutsche Quarzschmelze); CGW = Corning Glass Works; and d = doped.

nearly all high-volume manufacturers of natural vitreous silica probably use some form of sand beneficiation.

Summary

The unique properties of high-silica glasses define not only commercial applications, but also manufacturing difficulties. Following a period of extensive laboratory tinkering during the nineteenth century, commercially available products distinguished as clear (fused quartz) or opaque (fused silica) were developed at the beginning of the twentieth century, when many new mechanical techniques were invented.

Mechanical advances culminated with the development of a continuous-draw furnace in 1939. Chemical-processing advances (synthetic capability, raw-material beneficiation, and doping) permit property tailoring and blur the clear/opaque distinction. Today, the high-silica system comprises a family of glasses whose property range is determined by trace impurities or intentional dopants. Cation impurities are typically determined by raw materials and, generally, decrease in the order of sand > crystal $\approx$ beneficiated sand > synthetically derived raw materials. In general, the presence of metallic impurities will decrease optical transmission, increase the darkening tendency, and have relatively minor effects on thermal expansion and viscosity.

Anion impurities, principally OH^- and Cl^-, are primarily determined by the fusion technique used, and they effectively increase ir absorption and decrease darkening rate and viscosity. There are "wet" and "dry" processes for both the flame and electric-fusion techniques. Table IV categorizes some currently available commercial examples of the high-silica glass family according to raw-material type and fusion method.

Present and future development, driven by semiconductor- and fiber optic-industry demands, aims primarily at ultrahigh-purity, defect-free materials derived from synthetic, hybrid-synthetic, and sol/gel approaches.

Acknowledgments

The author would like to acknowledge the kind assistance of A. Brown, R. K. Cummins, and G. E. Dogunke of The General Electric Company; D. Kempe and M. Robinson of TAFQ; D. Rickard of Quartz Products Corp.; H. Nagashima of Toshiba; A. C. Kreutzer of Heraeus Amersil; T. Larkin of GTE-Sylvania; and L. Clark and E. T. Decker of Corning Glass Works.

References

[1] R. R. Sosman, The Properties of Silica; p. 96. The Chemical Catalog, New York, 1927.
[2] R. R. Sosman, The Properties of Silica; p. 96. The Chemical Catalog, New York, 1927.
[3] R. R. Sosman, The Properties of Silica; p. 96. The Chemical Catalog, New York, 1927.
[4] R. R. Sosman, The Properties of Silica; p. 96, 809. The Chemical Catalog, New York, 1927.
[5] R. R. Sosman, The Properties of Silica; p. 97. The Chemical Catalog, New York, 1927.
[6] R. R. Sosman, The Properties of Silica; p. 809. The Chemical Catalog, New York, 1927.
[7] R. R. Sosman, The Properties of Silica; p. 809. The Chemical Catalog, New York, 1927.
[8] R. R. Sosman, The Properties of Silica; p. 809. The Chemical Catalog, New York, 1927.
[9] R. R. Sosman, The Properties of Silica; p. 809. The Chemical Catalog, New York, 1927.
[10] R. R. Sosman, The Properties of Silica; p. 807. The Chemical Catalog, New York, 1927.
[11] R. R. Sosman, The Properties of Silica; p. 809. The Chemical Catalog, New York, 1927.
[12] R. R. Sosman, The Properties of Silica; p. 809. The Chemical Catalog, New York, 1927.
[13] G. Hetherington, Portrait of a Company (Wallsend: Thermal Syndicate, Ltd., 1906–1981); p. 3. Publisher?, 1981.
[14] G. Hetherington, Portrait of a Company (Wallsend: Thermal Syndicate, Ltd., 1906–1981); p. 3. Publisher?, 1981.
[15] G. Hetherington, Portrait of a Company (Wallsend: Thermal Syndicate, Ltd., 1906–1981); p. 3. Publisher?, 1981.

[16] R. R. Sosman, The Properties of Silica; p. 810. The Chemical Catalog, New York, 1927.
[17] G. Hetherington, Portrait of a Company (Wallsend: Thermal Syndicate, Ltd., 1906–1981); p. 20–21. Publisher?, 1981.
[18] G. Hetherington, Portrait of a Company (Wallsend: Thermal Syndicate, Ltd., 1906–1981); p. 29. Publisher?, 1981.
[19] J. A. Heany, U.S. Pat. No. 2 268 589, January 6, 1942.
[20] J. F. Hyde, U.S. Pat. No. 2 272 342, February 10, 1942.
[21] L. Clark; private communication, August 1984.
[22] M. E. Nordberg, U.S. Pat. No. 2 326 059, August 3, 1943.
[23] J. A. Winterburn, U.S. Pat. No. 3 275 408, September 27, 1966.
[24] H. P. Hood and M. E. Nordberg, U.S. Pat. No. 2 106 744, February 1, 1938.
[25] H. P. Hood and M. E. Nordberg, U.S. Pat. No. 2 315 329, March 30, 1943.
[26] R. Cummins; private communication, July 1984.
[27] W. Hanlein, U.S. Pat. No. 2 155 131, April 18, 1939.
[28] G. Hetherington, Portrait of a Company (Wallsend: Thermal Syndicate, Ltd., 1906–1981); Publisher?, 1981.
[29] G. E. Dogunke; private communication, October 1984.

Aluminosilicate Glasses

WILLIAM H. DUMBAUGH AND PAUL S. DANIELSON

Corning Glass Works
Corning, NY   14831

Composition, structure, and distinguishing properties of aluminosilicate glasses are discussed in relation to their commercial applications. Alkali aluminosilicates have been used in such products as aircraft windows, chemical apparatus, glass electrodes, etc. Alkaline earth aluminosilicates have found application as fibers, lamp envelopes, cooking utensils, electronic substrates, infrared domes, etc.

Aluminosilicate glass may be defined compositionally as simply a silicate glass containing alumina as a major constituent; to distinguish it from a borosilicate glass that contains alumina, the aluminosilicate has more alumina than boron on a molar basis. From a properties standpoint, aluminosilicates are generally the most refractory of glasses containing alkali and/or alkaline-earth modifiers. Their distinguishing feature is a steep viscosity–temperature curve (Fig. 1), which results in relatively high annealing and strain points, and yet the high-temperature viscosity is low enough to permit melting to good quality in conventional furnaces. Alkali aluminosilicates have high ion mobility, lending themselves to unique applications. Alkaline-earth aluminosilicates have tight structures, giving high electrical resistivities and low gas permeation.

Binary Aluminosilicate Glasses

The aluminosilicate binary is important in ceramic technology and has probably been studied more than any other system. However, it yields no glasses of commercial importance, even though the General Electric Company Ltd. obtained a British patent in 1937 on the eutectic glass at 5.5 wt% alumina.[1] In addition to high viscosity, the binary has a high liquidus temperature, and there is a large area of metastability, as shown in Fig. 2.[2-4] Forming a large piece of homogeneous glass at any appreciable alumina level is virtually impossible.

Glass can be made by extremely fast quenching, as shown in Table I. MacDowell and Beall state that glass stability decreases rapidly as alumina content increases, but that if modifying ions such as alkali or alkaline earth are added to the binary, "melts containing up to 50 mol% alumina could be easily quenched to homogeneous glasses".[4]

Alkali Aluminosilicate Glasses

Glass-Forming Areas

The approximate glass-forming regions for the Li, Na, and K aluminosilicate systems, as determined by Imaoka and Yamazaki,[9] are shown in Fig. 3. Imaoka and Yamazaki used 3–8-cm^3 Pt–Rh crucibles containing melts of $\approx$1 g, which were removed from the furnace and air-cooled. Different melt temperatures and

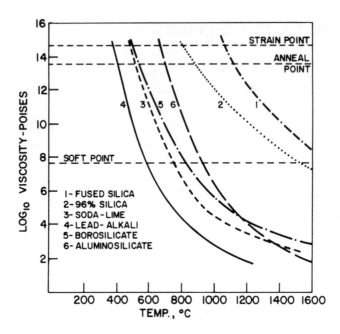

Fig. 1.   Viscosity curves of representative glasses.

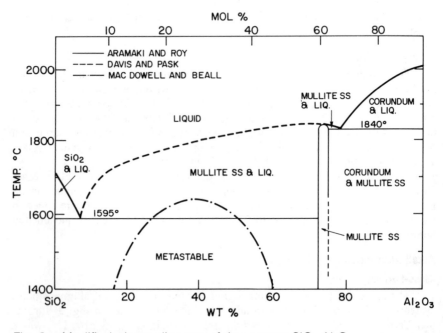

Fig. 2.   Modified phase diagram of the system SiO_2-Al_2O_3.

Table I. Al$_2$O$_3$-SiO$_2$, Upper Al$_2$O$_3$ Level to Form Glass

Wt%	Estimated Cooling Rate, °C/s	Sample Volume, cm^3	Method	Ref.
45	10^3	10^{-3}	Wire	5
53	5×10^3	3.4×10^{-5}	Wire	6
78	10^5	1.4×10^{-8}	Flame Spherulization	2
86	10^7	1.4×10^{-8}	Flame Spherulization	7
85	10^4	7.9×10^{-11}	Plasma (predicted)	8

cooling rates might be expected to change the glass-forming boundaries somewhat. No data are available for the Rb and Cs systems.

The lithium aluminosilicate system is best noted for the important group of glass-ceramics which may be crystallized from many of its glasses. Most of the literature describing this system, therefore, concentrates on nucleation and crystallization kinetics and the chemical and physical properties of the crystallized materials.

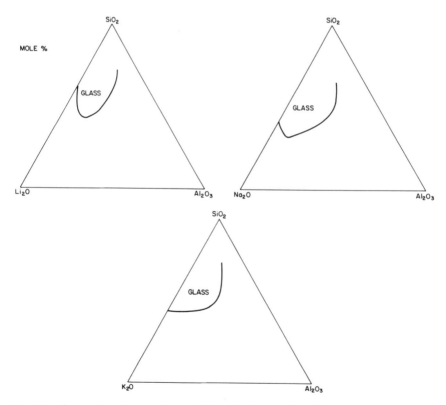

Fig. 3. Glass-forming regions of the alkali aluminosilicate systems.

Table II. Properties Sensitive to Al/M Ratio

Properties	Ref.
Refractive index	10
Density	10
Activation energy for viscous flow	12
Internal friction	18
Annealing point	
Strain point	
ir spectrum	10
uv cutoff	19
Knoop hardness	20
Electrical conductivity	20
Microwave dielectric loss	21
Activation energy for alkali diffusion	22
Ionic diffusion coefficients	22
Ionic selectivity (electrodes)	23
Shear and Young's moduli	24
Others	

The potassium aluminosilicate system has not received much attention in the literature; nor has it yielded significant commercial materials, with the exception of some electrode glasses.

The sodium aluminosilicate system has been the object of considerable fundamental study, as well as the basis for several important classes of commercial glass. Of the fundamental investigations, most work appears to have addressed the role of the aluminum ion in the glass structure.[10-17]

Structural Studies

The parameter of interest for most structure-related studies has been Al/M, the ratio of Al to alkali ions (or its reciprocal, sometimes called γ). Properties that have been reported sensitive to the Al/M ratio are included in Table II.

While there is still some disagreement, the following summary is a fair representation of current thinking on the structural role of aluminum. When the aluminum-to-alkali (Al/M) ratio is significantly less than 1, nearly all of the Al^{3+} ions are four-coordinated, occupying a "silicalike" network site with a negative charge. Presumably, a monovalent cation resides somewhere in the vicinity to provide charge neutrality. In principle, each Al ion added to an alkali silicate glass would replace a nonbridging oxygen (NBO) in the network until all NBOs were gone. In fact, the process may not be 100% complete. Thus, when Al/M is 1, most Al^{3+} ions are four-coordinated, and the number of NBOs is relatively small (probably a minimum).

When Al/M is greater than 1, some of the excess Al ions may be in sites of higher coordination.[16] Earlier work[11] indicated that excess Al ions were in octahedral sites, based on X-ray fluorescence peak shifts for Al. More recent studies, however, have failed to show significant peak variation with Al/M.[14] Nevertheless, since so many physical properties change with the Al/M ratio, it is tempting to conclude that changes in glass structure must be involved. Newer spectroscopic methods may eventually improve our understanding of such changes.

118

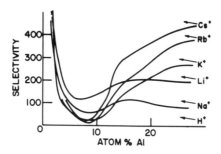

Fig. 4.   Selectivity of electrode glasses.

Commercial Alkali Aluminosilicate Glasses

Key features of commercially important glasses based on the alkali aluminosilicates include the variation of properties with Al/M, as already described, and an emphasis on those properties related to alkali mobility.[18-22]

One area of application for simple ternary glasses is that of ion-sensitive electrodes. Figure 4[23] shows how the selectivity order changes as a function of at.% Al for the sodium aluminosilicate glass system. Other ternary electrode glasses are included in Table III. A review of the electrochemistry of cation-sensitive glass electrodes has been written by Eisenman.[23] Doremus[25] has summarized selectivity theories and potentials of glass electrodes.

A second important group of applications for alkali aluminosilicate glasses results from their ability to be strengthened (or "chemically tempered") by an ion-exchange treatment below the strain point. Figure 5 shows diagrammatically how compressive surface stresses are developed through the replacement of smaller alkali ions in the glass by larger ions from the fused salt melt ("stuffing"). High strengths can be achieved using this process, as can fairly high central tensile stresses, leading to the desirable "dicing" of frangible materials. Table IV includes some commercial glasses which have been used in ion-exchange-strengthened products, along with the typical modulus of rupture values obtainable. The commercial glasses are no longer simple ternary compositions, but have been modified for manufacturability, color, and physical and chemical properties. Among the applications of ion-exchanged glasses are auto and aircraft windshields, frangibles, Maverick missile dome covers, tape reel flanges, electronic housings, aircraft mirrors, pipettes, and centrifuge tubes.

The chemical durability of glasses, based on the alkali aluminosilicate mineral nepheline syenite, has been used to advantage in designing host materials for

Table III.   Electrode Glasses

Mol%	NAS 11–18	KAS 20–5	NAS 27–4	LAS 15–25
Li_2O	0	0	0	15
Na_2O	11	0	27	0
K_2O	0	20	0	0
Al_2O_3	18	5	4	25
SiO_2	71	75	69	60
Selects	Na	K	K	Li

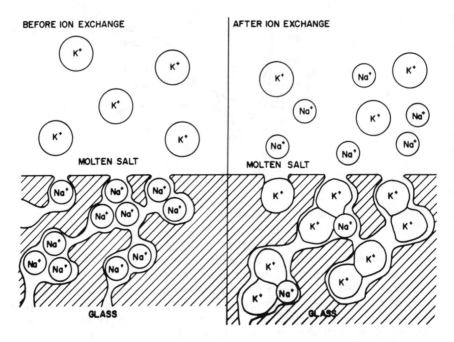

BEFORE ION EXCHANGE

AFTER ION EXCHANGE

MOLTEN SALT

MOLTEN SALT

GLASS

GLASS

Fig. 5. Ion-exchange strengthening.

nuclear waste fixation. Table V shows the compositions of two glasses developed and studied by researchers for Atomic Energy of Canada Ltd.[26,27] Samples of the second glass (15/85), containing fission products, have been on test in the ground

Table IV. Ion-Exchange-Strengthened Glasses

	Code 0317	Code 0331	Code 0417	Code PPG–IX
		Wt%		
SiO_2	61.4	65	62.4	+
Na_2O	12.7	8.7	13.4	+
K_2O	3.6	0.1	3.0	
MgO	3.7	0.8	4.1	ZrO_2
CaO	0.2	0	0.5	
Al_2O_3	16.8	21	15.3	+
TiO_2	0.8	0 FeO	0.4	
Li_2O	0	3.8	0	+
Soft. pt., °C	871	800	826	−
Anneal. pt., °C	624	551	585	−
Strain pt., °C	574	505	538	−
Exp. coeff. (25°–300°C), $\times 10^7/°C$	87	76	89	−
MOR { MPa	551	241	345	−
MOR { kpsi	80	35	50	−

Table V. Nuclear-Waste-Fixation Glasses*

Chalk River (Canada) Glass 843, wt%	
Na$_2$O	9
CaO	12
Al$_2$O$_3$	24
SiO$_2$	55
Lime-Nepheline Syenite Glass (15/85), wt%	
Na$_2$O + K$_2$O	14
MgO + CaO	14.6
Al$_2$O$_3$	19.2
SiO$_2$	46.7
Fission products	5.5

*Ref. 26.

at Chalk River Nuclear Lab for 20 years, and have shown excellent long-term durability and low leach rates.

Alkaline-Earth Aluminosilicate Glasses

The glass-forming areas for alkaline-earth aluminosilicate glasses, as determined by Imaoka and Yamazaki,[9] are shown in Fig. 6. The experimental techniques were the same as described for the alkali aluminosilicates. The calcium glass region is the largest, provides the most stable glasses, and is the most important commercially.

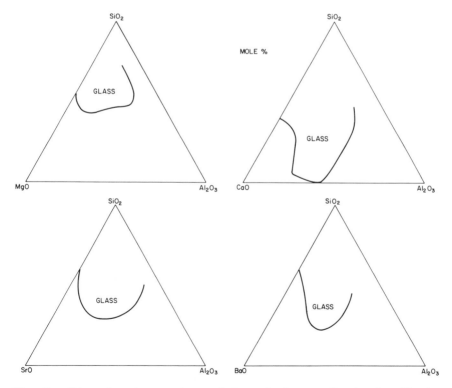

Fig. 6. Glass-forming regions of the alkaline-earth aluminosilicate systems.

121

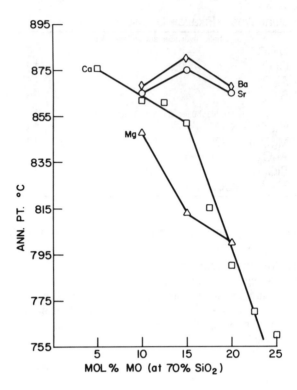

Fig. 7.   Effect of alkaline-earth-oxide-to-alumina ratio on annealing point.

The following data characterizing these glasses were obtained on 300- to 1000-g melts in Pt–Rh crucibles. The more refractory glasses, e.g., those above ≈70 mol% SiO_2, were melted in a gas-oxygen-fired furnace at 1800°C, and the others were melted in a resistance-heated furnace at 1600°C. The melts were cast into a slab and annealed at the appropriate temperature.

The behavior of aluminum in the structure of alkaline-earth glasses is similar to that described for the alkalis.[28] The effect of changing aluminum coordination as the alkaline-earth-oxide-to-alumina ratio is varied can be seen by its influence on annealing point in Fig. 7. Some glasses in the MgO system, particularly toward high alumina, tend to phase-separate with annealing. This produces anomalously high annealing points. Expansion coefficients (Fig. 8) for the same series do not exhibit this effect, but increase monotonically with increasing alkaline earth.

When the molar ratio of alkaline-earth oxide to alumina is 1:1, the effect of silica level on annealing point is as shown in Fig. 9, and the effect on expansion coefficient is as in Fig. 10.

Since the calcium aluminosilicate system is the basis for numerous commercial compositions, it has been explored in more detail by characterizing eutectic glasses. A eutectic composition is a good starting point for deriving a commercial glass, since liquidus viscosity is very important for many forming processes. Figure 11[29] shows the location of these compositions, numbered in accordance with the original work of Rankin and Wright.[30] Properties are listed in Table VI. Eutectic 2 has been the base for such glasses as Owens-Corning E-glass fiber,[31] tungsten-

122

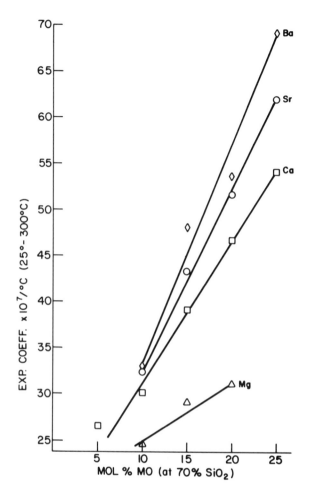

Fig. 8. Effect of alkaline-earth-oxide-to-alumina ratio on expansion coefficient.

halogen-lamp envelopes,[32,33] and the cladding glass for Corelle* dinnerware.[34]

The dc resistivity of calcium aluminosilicate glasses generally decreases with increasing silica (Fig. 12), based on a random collection of data from all measurements on glasses in this system. The only significance of the straight line is to indicate the resistivity trend over a sizeable range of silica. The same kind of plot (Fig. 13) shows the effect of silica level on infrared cutoff. Eutectic 7 has an optimum combination of infrared transmission, high mechanical hardness, moderate thermal expansion, and good weatherability for application as infrared missile domes. Corning Code 9753 glass[†] has the eutectic composition and has been used in the Redeye missile. Schott has a similar glass called IRG N6.[35] Eutectic 14 has better infrared transmission than eutectic 7, but poorer physical and chemical

---

*Corning Glass Works, Corning, NY.
[†]Corning Glass Works.

123

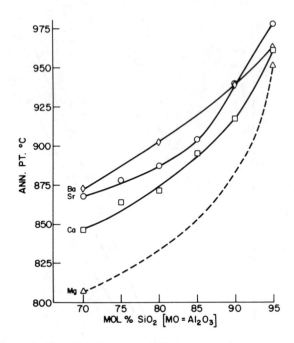

Fig. 9.   Effect of silica level on annealing point.

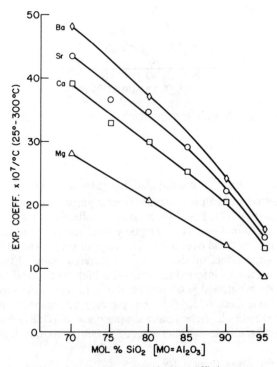

Fig. 10.   Effect of silica level on expansion coefficient.

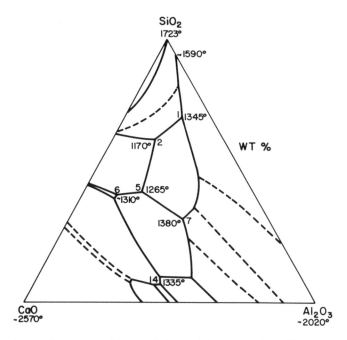

Fig. 11.   Eutectic compositions of the calcium aluminosilicate system.

properties. Barr and Stroud manufactures a glass based on a modification of eutectic 14, called BS 37A, for use as an infrared transmitting window.[36]

Modification of a eutectic composition often involves using more than one alkaline earth to achieve the desired properties and, hopefully, lower the liquidus even more, for improved stability. Since the aluminosilicates can be difficult to melt, e.g., eutectic 1 requires ≈1800°C and eutectic 2, 1550°C, boric oxide is sometimes added to improve meltability without seriously affecting desired properties. An example of one way to incorporate boric oxide is given in Table VII.

The earliest applications for alkaline-earth aluminosilicate glasses were combustion tubes for organic analysis and envelopes for discharge lamps, an application seldom used today. Over the years, Schott has had a series of glasses called Supremax[37]; they may well have been the first of this type and are still available. Actually, on March 10, 1925, Fred M. Locke and Fred J. Locke of Victor, New York, obtained a U.S. patent that was broad enough to include almost every glass mentioned in this paper.[38] Selected glasses originally developed for these applications are shown in Table VIII. They have also been used for other products; e.g., the Corning Code 1720 composition[‡] was used for consumer top-of-stove ware such as coffee-percolator bodies. They have also been useful for articles requiring glass-to-metal seals: In general, glasses with expansions below $40 \times 10^{-7}/°C$ will seal to tungsten, and those in the low 40s will seal to molybdenum.

One presently active application is envelopes for tungsten-halogen lamps. Some of these glasses are listed in Table IX. Owens-Illinois EE–2 glass[§] is directed to electron-tube envelopes.

[‡]Corning Glass Works.
[§]Owens Illinois Inc., Toledo, OH.

125

Table VI. Eutectic Glasses of the Calcium Aluminosilicate System

Eutectic No.		1	2	5	6	7	14
Wt% {SiO$_2$		70.4	62.1	42.0	41.0	31.7	6.95
{Al$_2$O$_3$		19.8	14.6	20.0	11.8	29.1	43.35
{CaO		9.8	23.3	38.0	47.2	29.1	49.70
Exp. coeff. (25°–300°C), ×10^7/°C		33.1	54.7	76.5	88.7	59.5	84
Anneal. pt., °C		883	772	781	770	832	822
Strain pt., °C		831	728	746	738	800	≈796
Temp. of 1000 poises, °C		1612	1323	1166		1230	
log ρ (Ω·cm) {250°C		12.7	14.2	16.1	14.8	18.0	15.8
{350°C		10.6	11.8	13.3	12.3	15.0	13.0
Dielectric const., 10 kHz, 25°C		6.3	7.1	9.3	9.9	8.9	12.5
Polished plate durability (95°C) —wt. loss, mg/cm²	5% HCl, 24 h	0.04	0.04	High	138	Dissolves	157
	5% NaOH 6 h	0.82	0.96	0.76			1.2

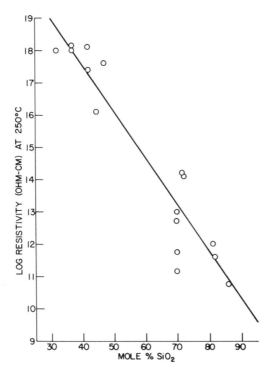

Fig. 12. Log resistivity vs silica content for calcium aluminosilicate glasses.

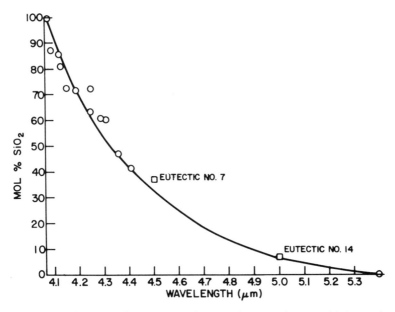

Fig. 13. Wavelength where transmittance is 50% (2 mm thickness) at ir absorption edge vs silica content.

Table VII.   Effect of Boric Oxide in Eutectic 2
(SiO_2, 64.9 mol%, Al_2O_3, 9.1 mol%, CaO, 26.0 mol%)

Mol% Substitution B_2O_3 for SiO_2	Exp. Coeff. (25°–300°C) $\times 10^7/°C$	Anneal Pt., °C	T_{1000P}, °C	5% HCl–95°C–24 h (Wt. loss, mg/cm^2)
Base	55	772	1336	0.04
1	55	752	1310	0.06
2	55	746	1293	0.08
3	55	730	1242	
7	57	658	1214	0.17

Table VIII.   Early Commercial Alkaline-Earth Aluminosilicate Glasses

	Supremax* (Schott)	H.26 (G.E.C.)	C37 (A.E.I.)	No. 40–Mazda (U.S.S.R.)	742a (Osram)	Code 1720 (Corning)
			wt%			
SiO_2	53.5–56.4	54	55.8	57.6	51.3	62
Al_2O_3	20.0–23.5	21	23.0	25	25.3	17
B_2O_3	7.0–10.0	8	5.1		1.0	5
MgO	8.9–10.0			8	4.2	7
CaO	4.6–15.2	14	13.0	7.4	8.3	8
BaO		3	3.1		5.3	
Na_2O	0.7–2.0					1
K_2O				2		
P_2O_5					4.6	
Exp. coeff., $\times 10^7/°C$	41	43	42.5	37.5	31	42
Range of exp., °C	20–300	20–350	50–400	20–100	20–300	0–300
Soft. pt., °C	950					915
Anneal. pt., °C	740	≈720	≈760	≈806	Tg = 735	712
Strain pt., °C	690		700			667
Density, g/cm^3	2.56		2.55	2.56	2.62	2.52
Temp. for $\rho = 10^8$ $\Omega \cdot$cm, °C	616			400	450	463
Ref.	37, 39	40	41	42	43	44

*There are a number of Supremax glasses; properties are for 8409 (Ref. 39).

Another growing market is for substrates which must be flat, smooth, electrically inactive, and have thermal expansion compatible with the deposited material. Examples are tabulated in Table X.

Acknowledgments
The authors are grateful to Corning Glass Works and the companies listed in Ref. 53 for technical information contained in this paper.

Table IX. Envelopes

	Corning		G.E.		O-I EE-2	Philips 18	Schott 8252	Toshiba HA-4
	Code 1724	Code 1725	177	180				
Soft. pt., °C	923	≈1016	1130	1020	960	955.65	950	971
Anneal. pt., °C	717	790	865	805	765	740	($T_g = 790$)	($T_g = 770$)
Strain pt., °C	672	732	805	755	720	710		
Exp. Coeff. (0°–300°C), ×10^7/°C	43.5	43.7	38	43	43	38.5 (30°–300°C)	50.3 (100°–300°C)	46 (100°–300°C)
Density, g/cm^3	2.63	2.56	2.70	2.68	2.59	2.55	2.65	2.70
log ρ (Ω·cm), 350°C	12.4		10.5	11.1	11.7	10.0		
Ref.	33	45	46	46	47	48	49	50

Table X. Substrates

	Corning		Hoya		O–I		Schott
	Code 1723	Code 1717	Na-40	LE30	EE5	EE9	S-8020
Exp. coeff. (0°–300°C), $\times 10^7$/°C	46	38	43 (100°–300°C)	38	31	32	36 (20°–300°C)
Working pt., °C	1168	1479		921	1395	1305	1197
Soft. pt., °C	908		($T_g = 730$)	686	1070	1020	
Anneal. pt., °C	710	854	656	625	819	809	($T_g = 679$)
Strain pt., °C	665	799			772	756	
Density, g/cm³	2.64	2.55	2.87	2.58	2.57	2.54	2.83
Young's mod., kg/mm²	8789	8437	9420	7540			8137
log ρ (Ω·cm), 350°C	11.3	10.1			11.6	11.3	
Application	Resistor cane	Si substrate	Substrate for displays	Substrates, liquid level gauge	Seal to Si	RF sputtering of thin films	Photomask, mirrors
Ref.	44		51	51	47	47	52

130

References

[1]General Electric Company Ltd., "Improvements in or Relating to Intermediate Glasses for Sealing Conductors into Quartz Vessels," Br. Pat. No. 463 889, 1937.

[2]S. Aramaki and R. Roy, "Revised Phase Diagram for the System Al_2O_3-SiO_2," J. Am. Ceram. Soc., 45 [5] 229–42 (1962).

[3]R. F. Davis and J. A. Pask, "Diffusion and Reaction Studies in the System Al_2O_3-SiO_2," J. Am. Ceram. Soc., 55 [10] 525–31 (1972).

[4]J. F. MacDowell and G. H. Beall, "Immiscibility and Crystallization in Al_2O_3-SiO_2 Glasses," J. Am. Ceram. Soc., 52 [1] 17–25 (1969).

[5]M. S. R. Heynes and H. Rawson, "An Investigation into Some High Melting Point Glass Systems and their Use as Solvents for Uranium Dioxide," Phys. Chem. Glasses, 2 [1] 1–11 (1961).

[6]D. J. Thorne, "Glass Formation in Refractory Oxide System Based on Alumina," Proc. Brit. Ceram. Soc., 14, 131–45 (1969).

[7]T. Takamori and R. Roy, "Rapid Crystallization of SiO_2-Al_2O_3 Glasses," J. Am. Ceram. Soc., 56 [12] 639–44 (1973).

[8]M. S. J. Gani and R. McPherson, "Glass Formation and Phase Transformations in Plasma-Prepared Al_2O_3-SiO_2 Powders," J. Mater. Sci., 12 [5] 999–1009 (1977).

[9]M. Imaoka and T. Yamazaki, Report of the Institute of Industrial Science, The University of Tokyo, 18 [14] (Serial No. 118) 15–17 (1968).

[10]D. E. Day and G. C. Rindone, "Properties of Soda Aluminosilicate Glasses: I, Refractive Index, Density, Molar Refractivity, and Infrared Absorption Spectra," J. Am. Ceram. Soc., 45 [10] 489–96 (1962).

[11]D. E. Day and G. C. Rindone, "Properties of Soda Aluminosilicate Glasses: III, Coordination of Aluminum Ions," J. Am. Ceram. Soc., 45 [12] 579–81 (1962).

[12]E. F. Riebling, "Structure of Sodium Aluminosilicate Melts Containing at least 50 mol% SiO_2 at 1500°C," J. Chem. Phys., 44 [8] 2857–65 (1966).

[13]B. E. Yoldas, "Coexistence of Four- and Six-Co-ordinated Al^{3+} In Glass," Phys. Chem. Glasses, 12 [1] 28–32 (1971).

[14]S. Sakka, "Study of the Coordination Number of Aluminum Ions in Aluminosilicate Glasses by Means of $AlK\alpha$ Fluorescence X-ray Spectra," Yogyo-Kyokai-Shi, 85 [4] 168–73 (1977).

[15]K. Hunold and R. Bruckner, "Physical Properties Structural Details of Sodium Aluminosilicate Glasses and Melts," Glastech. Ber., 53 [6] 149–61 (1980).

[16]A. Klonkowski, "The Structure of Sodium Aluminosilicate Glass," Phys. Chem. Glasses, 24 [6] 166–71 (1983).

[17]J. Wong and A. Angel, Glass Structure by Spectroscopy; p. 134 and references therein. Marcel Dekker, New York, 1976.

[18]D. E. Day and G. E. Rindone, "Properties of Soda Aluminosilicate Glasses: II, Internal Friction," J. Am. Ceram. Soc., 45 [10] 496–504 (1962).

[19]J. Wong and A. Angel, Glass Structure by Spectroscopy; p. 166 and references therein. Marcel Dekker, New York, 1976.

[20]R. Terai, "Self-Diffusion of Sodium Ions and Electrical Conductivity in Sodium Aluminosilicate Glasses," Phys. Chem. Glasses, 10 [4] 146–52 (1969).

[21]J. A. Topping and J. O. Isard, "The Dielectric Properties of Sodium Aluminosilicate Glasses at Microwave Frequencies," Phys. Chem. Glasses, 12 [6] 145–51 (1971).

[22]E. L. Williams and R. W. Heckman, "Sodium Diffusion in Soda-Lime-Aluminosilicate Glasses," Phys. Chem. Glasses, 5 [6] 166–71 (1964).

[23]G. Eisenman, "The Electrochemistry of Cation-Sensitive Glass Electrodes"; p. 223 in Advances in Analytical Chemistry and Instrumentation, Vol. 4. Edited by C. N. Reilly. Wiley & Sons, New York, 1965.

[24]R. J. Eagan and J. C. Swearengen, "Effect of Composition on the Mechanical Properties of Aluminosilicate and Borosilicate Glasses," J. Am. Ceram. Soc., 61 [1–2] 27–30 (1978).

[25]R. H. Doremus, "Ion Exchange and Potentials of Glass Electrodes"; pp. 253–80 in Glass Science. Wiley & Sons, New York, 1975.

[26]K. B. Harvey and C. D. Litke, "Model for Leaching Behavior of Aluminosilicate Glasses Developed as Matrices for Immobilizing High-Level Wastes," J. Am. Ceram. Soc., 67 [8] 553–56 (1984).

[27]J. C. Tait and D. L. Mandolesi, "The Chemical Durability of Alkali Aluminosilicate Glasses," Rept. No. AECL–7803, Atomic Energy of Canada Ltd., 1983.

[28]M. Yamane and M. Okuyama, "Coordination Number of Aluminum Ions in Alkali-Free Alumino-Silicate Glasses," J. Non-Cryst. Solids, 52, 217–26 (1982).

[29]E. F. Osborn and A. Muan, "Phase Equilibrium Diagrams of Oxide Systems," Plate 1, revised and redrawn. The American Ceramic Society and the Edward Orton, Jr., Ceramic Foundation, 1960.

[30]G. A. Rankin and F. E. Wright, "The Ternary System CaO-Al_2O_3-SiO_2," Am. J. Sci. — Fourth Series, 39 [229] 1–79 (1915).

[31]P. F. Aubourg and W. W. Wolf, "Glass Fibers: Glass Composition Research"; these proceedings.

[32] W. H. Dumbaugh, "Glass Envelopes for Tungsten-Halogen Lamps," U.S. Pat. No. 4409337, October 11, 1983.

[33] W. H. Dumbaugh, "Envelopes for Tungsten-Halogen Lamps," U.S. Pat. No. 4394453, July 19, 1983.

[34] D. A. Duke, W. H. Dumbaugh, J. E. Flannery, J. W. Giffen, J. F. MacDowell, and J. E. Megles, "Glass Laminated Bodies Comprising a Tensilely Stressed Core and a Compressively Stressed Surface Layer Fused Thereto," U.S. Pat. No. 3673049, June 27, 1972.

[35] Schott Glaswerke, Infrared Transmitting Glasses, Product Information Sheet No. 3112/1e, May 1982.

[36] Barr and Stroud Ltd., "Calcium Aluminate Types BS37A and BS39B," Bull. No. 1554, Glasgow, Scotland, 1962.

[37] M. B. Volf, Technical Glasses; p. 419. Pitman, London, 1961.

[38] F. M. Locke and F. J. Locke, "Glass," U.S. Pat. No. 1529259, March 10, 1925.

[39] Schott Glaswerke, "Physical and Chemical Properties of Industrial Glasses," Brochure No. 2057/1e.

[40] M. B. Volf, Technical Glasses; p. 345. Pitman, London, 1961.

[41] M. B. Volf, Technical Glasses; Table IX. Pitman, London, 1961.

[42] M. B. Volf, Technical Glasses; p. 396. Pitman, London, 1961.

[43] M. B. Volf, Technical Glasses; p. 395. Pitman, London, 1961.

[44] D. C. Boyd and D. A. Thompson, Glass; pp. 807–80 in Kirk-Othmer: Encyclopedia of Chemical Technology, 3rd ed., Vol. 11. Wiley & Sons, New York, 1980.

[45] P. S. Danielson, "Glass Envelopes for Tungsten-Halogen Lamps," U.S. Pat. No. 4302250, November 24, 1981.

[46] General Electric, "Types 177 and 180 Glass Tubing," Product Data Sheet 7610-a, December 1, 1978.

[47] Owens-Illinois, "Miscellaneous Glasses," Product Data Sheet.

[48] M. B. Volf, Technical Glasses; Tables IX and X. Pitman, London, 1961.

[49] Schott Glaswerke, "Schott-Aluminosilicate Glass 8252 for Halogen-Lamps," Preliminary Product Information, 1980.

[50] Omori and Ichinose, "Lamps and Their Materials," Toshiba Review 37–2, 1982; p. 122.

[51] Hoya Optics, "Non-Alkali Glass NA–40" and "LE30 Low-Expansion, Heat-Resistant Glasses," Product Information Sheets.

[52] Schott Glaswerke, "Data Summary for S–8020," Information Sheet.

[53] Company Addresses:

Hoya.Optics, Inc., 3400 Edison Way, Fremont, CA 94538, (415)490-1880.

Owens-Illinois, Television Products Division, Technical Products, One SeaGate, Toledo, OH 43666, (419)247-5000.

JENAer Glaswerk Schott & Gen., Werk Landshut, Christoph-Dorner-Str. 29, Postfach 2520, D-8300 Landshut, FRG, tel. (0871)2071.

General Electric Lamp Components Sales Operation, 21800 Tungsten Road, Cleveland, OH 44117, (216)266-2451.

Schott Glaswerke, Optics Division, Sales Dept., Optical Glass, P.O. Box 2480, D-6500 Mainz, FRG, tel., (06131)661; telex, 4187922-0 smd.

Phase-Separated and Reconstructed Glasses

WOLFGANG K. HALLER

4620 North Park Ave.
Chevy Chase, MD   20815

The history of borate glasses, observation of thermally induced leachability, and the technological development of reconstructed high-silica glass is described. Metastable liquid/liquid immiscibility as the cause of leachability is reviewed.

Vycor* high-silica glass is a reconstructed glass that has obtained major commercial significance, and therefore this paper deals mainly with the historical developments, properties, and uses of Vycor. Other commercial glasses in which phase separation plays an important role are opal glasses, glass-ceramics, and photosensitive glasses: They are covered elsewhere.

This mainly technological review will cover the field of phase-separated and reconstructed glass from its beginnings to the full commercialization of Vycor, and will mention some improved and related products. The review will end around 1960, and work done in the 1960s and 1970s on the more esoteric points of nucleation, growth, metastable liquid/liquid-phase separation, spinodal decomposition, and ripening of phase-separated structures will not be covered here or will be mentioned only briefly.

Borate Glasses

The origins of reconstructed glasses are intimately associated with the use of B_2O_3 in silicate glasses. The first deliberate use of B_2O_3 was by Otto Schott in the 1880s, when he tried to modify the optical properties of the then-available alkali-alkaline earth silicate and lead-silicate glasses. Among the many oxides that he explored, he found borates particularly useful as a flux for low-refractive-index glasses. These optical borate glasses, although very important for optical scientific instruments such as microscopes, telescopes, and cameras, served a strictly low-tonnage, high-value industry.

At almost the same time, the Industrial Revolution brought coal gas to the cities, replacing kerosene as an energy source for home and street lighting. The most effective lamp consisted of an emitter called the Auer von Welsbach mantle, which was heated by a high-temperature gas flame. The emitter was surrounded by a glass cylinder which, because of the high temperature involved, was subject to frequent thermal fracture.

It was Otto Schott, again, who remembered the good thermal-shock resistance that borates impart to silicate glasses; by producing borate glass-lamp cylinders, he initiated the first large-volume production of borate glasses. According to contemporary sources, the lamp glass contained 24 wt% B_2O_3: This must have made the

---

*Corning Glass Works, Corning, NY.

glass rather low in hydrolytic durability and very high-priced, considering that boric acid was, at that time, an expensive raw material.

The next quantum step in borate use occurred when borosilicate glassware was introduced into the chemical laboratory and the kitchen. These developments were made possible, however, only by the discovery of extensive borate deposits in the American West around the turn of the century. In 1915 and 1919, Sullivan and Taylor[1] of Corning Glass Works filed patents for a borosilicate glass of which Pyrex 7740[†] later became the best-known representative. This was followed in Germany, in 1920, by the Jena G20[2] glass, and by other similar glasses in Czechoslovakia and England. Although Pyrex 7740, with a B_2O_3 content of about 13% and an expansion coefficient of 33×10^{-7}, soon established itself in the United States, other countries settled for lower B_2O_3 contents and slightly higher expansion coefficients for their kitchen and laboratory ware. Some claims made for those glasses were better alkali resistance and ability to seal to molybdenum, but their main advantage was lower price, since B_2O_3 had to be imported.

Since chemical durability is routinely tested in the production of laboratory ware, early reports mention unexpected catastrophic chemical-durability loss caused by composition fluctuations during fabrication. Such extreme changes of chemical durability with rather minor changes in composition were unexpected, but at that stage, no unusual phenomena were suspected as the cause.

Leachable Glass

In 1926, Turner and Winks[3] studied systematic B_2O_3 substitution in soda-lime-silica glasses and reported glasses with high weight loss on acid extraction if B_2O_3 was increased above 30%. They also examined the leftovers of their extraction experiments and found a porous, rigid, silica-gel-like body that retained the shape of the original glass.

At that stage, glass chemists realized that they were dealing with a most uncommon phenomenon. There was ample speculation about the mechanism underlying this selective leachability, but two scientists at Corning Glass Works, Hood and Nordberg,[4] succeeded in turning an oddity and scientific puzzle into an immensely useful process and product: They found that objects made from certain alkali borosilicate glasses, after conversion into porous objects by acid leaching, could be subsequently sintered into reconstructed, pore-free, high-silica glass.

The discoverers of this process called the end-product Vycor. The ingenuity and hard work that made such an unlikely process succeed was a true milestone in glass technology.

Vycor 96% Silica Glass

Vycor glass is almost pure vitreous SiO_2 with all the desirable basic properties of fused quartz. It has a low expansion coefficient, a high deformation temperature, and excellent chemical durability and uv transmittance. However, although fused quartz requires melting temperatures of 1900°C, the borosilicate glass precursor of Vycor can be melted some 400°C lower, and reconstruction or sintering takes place another 500°C lower. Table I shows the temperatures involved in making fused quartz and Vycor glass; also included are Pyrex 7740 and the softening points of the glasses, for comparison. From 1938 on, Nordberg and Hood filed a large number of patents, first for improvements of the process to make the new high-silicate glass, later for modifications to make other related products.

[†]Corning Glass Works.

134

Table I. Comparison of Processing Temperatures (°C)

	Melting	Sintering	Softening (10^{7.65} P)
Quartz	≈1900		≈1580
Borosilicate to			
make Vycor*	≈1500		
Reconstructed			
Vycor		900–1200	≈1500
Pyrex 7740†	≈1500		820

*Corning Glass Works.
†Corning Glass Works.

Table II. Properties of Vycor and Other Glasses

	Vycor	Fused Silica	Pyrex 7740	Soft Glass
Thermal expansion, $10^{-7}/°C$	8	8	33	92
Softening point, °C ($10^{7.5}$ P)	1500	1580	820	695
Strain point, °C ($10^{14.8}$ P)	820	890	515	470
Density, g/cm³	2.18	2.18	2.23	2.47
Refr. index, N_D	1.458	1.458	1.474	1.512

Figure 1 shows a schematic of the process, as described in later patents. The composition of the starting glass is quoted from one of the patents. In this preferred composition, heat treatment of the glass object is necessary to make it acid-leachable. Heat-treatment temperatures range from 500°–600°C and can last from several hours to three days. Sintering is done between 900° and 1200°C.

A major bottleneck of the process was a tendency toward fracture during leaching, due to swelling; several patents deal with special glass compositions minimizing this problem. The process is most useful for making thin-walled objects, because thick objects have a greater tendency to fracture during leaching, and because leaching and washing time also increases with thickness. Typical properties of Vycor are shown in Table II.

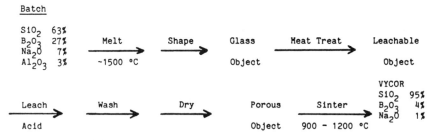

Fig. 1. Processing of Vycor glass.

135

Table III. Use-Related Properties of Vycor Glass

	Labware, Chemical Processing	Metallurgy	Glass-Melting Liners	Spacecraft Windows, Heat Lamps	uv Lamps	Radomes
Chemical resistance	X		X			
Low contamination	X		X			
High deformation temperature	X	X		X		X
Thermal-shock resistance		X		X		X
uv Transmission					X	
ir Transmission						X

New processes to improve Vycor were invented over the years. Elmer[5] worked on ways to reduce water content, thus improving infrared transmission. In one such process he treated the porous precursor with steam, driving out water with water. This sounds like a contradiction, but Vycor contains traces of B_2O_3 which volatilize as boric acid in the steam atmosphere. Apparently, the boric-oxide-free Vycor then has less tendency to retain water when dried and sintered.

Vycor is used in laboratory glassware as crucibles, dishes, and tubings to construct instruments. In many industrial processes Vycor is used for containers: For instance, large Vycor dishes are used for calcining of television-tube phosphors, and metallurgists use Vycor tubes to protect throwaway thermocouples that are pushed through the floating slag layer into the interior of molten metal. In metal casting, Vycor is used as a mold; it is also used to melt glasses by making it the liner of refractory crucibles. Because of its thermal-shock resistance and high deformation temperature, spacecraft windows are made of Vycor, as are envelopes for heat lamps and high-intensity lights. Since it has good ultraviolet transmission, it is used for gas-discharge lamps, particularly germicidal lamps. Dewatered Vycor transmits infrared rays and has been used for radomes of heat-seeking guided missiles. Table III summarizes uses of Vycor and the respective physical and chemical properties that lead to particular applications.

Modified Vycor Glass

Modified Vycor products have been suggested and made: In 1950, Lukas[6] suggested that objects with gradients of composition could be produced by impregnating the porous Vycor intermediate with successive solutions of carbonates, hydroxides, or nitrates of other elements, then drying and sintering the objects; such glasses were tailored for monolithically graded seals.

By sintering the porous precursor in an ammonia atmosphere, scientists at Corning produced partially nitrided silica, with increased working temperature[7]; these products were used for gas turbines.

Porous Glass

The leached, nonsintered, porous intermediate of Vycor production was nicknamed *Thirsty Glass,* because it sticks to the tongue. At first, it was treated more as a scientific curiosity than a commercial product with large-scale applications. It has a void volume of $\approx28\%$, a surface area of ≈300 m^2/g, and a pore distribution ranging from 3.0–4.5 nm (30–45 Å). It did, however, find a number of uses.

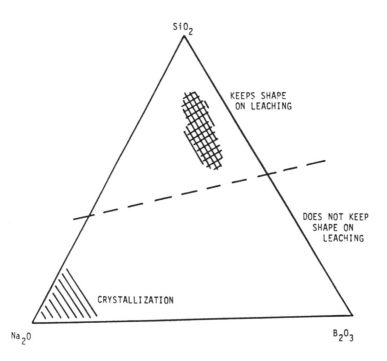

Fig. 2. Schematic of system Na_2O-B_2O_3-SiO_2.

As a nondusting drying agent or water getter, this intermediate product absorbs up to 15% of its weight in water and is less messy than silica gel, therefore lending itself particularly well to use in scientific instruments. It has been suggested as a virus and bacteria filter, as a catalyst carrier, and as a substrate for adsorption chromatography of low-molecular-weight substances.

Phase-Separation Mechanism

This paper has so far concentrated on the technological aspects and the history of phase-separated and reconstructed glasses. The phase-separated terminology, however, is more recent and suggests a particular mechanism not understood when the so-far-described technological developments took place.

The original base glass suggested for the manufacture of Vycor belongs to the system R_2O-B_2O_3-SiO_2, where R_2O is usually Na_2O but can also be K_2O or Li_2O. Figure 2 shows a schematic of this composition diagram. At the high-Na_2O corner, in the lower left, crystallization interferes with glass formation; otherwise, glasses could be made over the whole field. The glasses are subject to attack by acid over almost the whole field: Some compositions do this in the melt-quenched form; others have to be annealed or heat-treated to become acid-leachable. In the region below the dashed line, the glasses do not retain their shape with leaching; they decompose into a paste. The crosshatched region is the composition region patented by Hood and Nordberg,[4] where, in the Na_2O-containing system, a relatively short heat treatment suffices to induce acid leachability.

Adding other oxides reduces the trend toward acid leachability. Adding Al_2O_3, for instance, produces the compositions of borosilicates used for laboratory glassware such as Pyrex 7740, and adding Al_2O_3 and CaO or MgO produces the

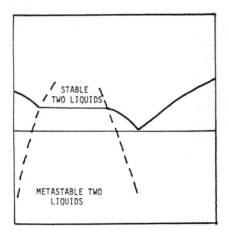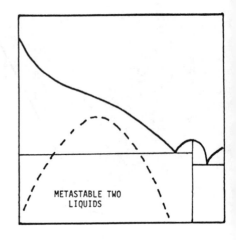

Fig. 3. Schematic of immiscibility in a binary system (from Ref. 12).

composition of Schott G20 or certain fiberglasses. Some of the technical boro-silicates mentioned actually still tend to become extractable with prolonged heat treatment[8,9] and certainly do so if, through production problems, they stray from their target composition.[2]

Early electron microscopy done in Germany by Skatulla et al.[10] showed that the leachable glasses were not homogeneous; this was also deduced from simple observation of opalescence, i.e., light scattering, in glasses that had been heat-treated for a long time.

Although everybody agreed on the principle of inhomogeneity, investigators had only a very vague idea about its causes and speculated clusters of ions or the formation of molecular groups similar to compounds. The decisive event that cleared the picture appears to have been a paper given by Roy and Ruiz-Menacho in Chicago in 1959.[11] In this paper they suggested a special case of liquid/liquid immiscibility, which they called metastable liquid immiscibility — in other words, the well-known phenomenon of a liquid mixture separating into two immiscible liquid phases, but in this case happening below the liquidus temperature. Because of the high viscosities of glass melts that prevail at subliquidus temperatures, phase separation does not lead rapidly to layers of liquids but rather to finely dispersed liquid phases. Figure 3 shows a graph from Roy's paper for a binary system[12]: To the left is the usual, easily observable immiscibility dome extended into the subliquidus region; to the right is a dome which is completely subliquidus. When a glass of this composition is quenched quickly, it becomes two-phase, but with very small domains that cause no light scattering. If, however, the glass is heat-treated for a long period, the domains will grow and manifest themselves by light scattering.

Roy[12] painted a very logical and (retrospectively) simple picture of this puzzling phenomenon. This was in 1959, 20 years after Hood and Nordberg[4] had succeeded in producing a commercial product using a newly discovered, rather mysterious phenomenon. The 1960s and 1970s saw a whole new wave of theoretical and experimental work dedicated to metastable subliquidus phase separation: Immiscibility domes were experimentally determined and their shapes were explained; the phenomena preceding phase separation were investigated; phase morphology and its development kinetics were studied. Then, after this outburst of papers, the publication rate in the field declined to a trickle. We are still far from a full understanding of the phenomena, but most of the champions of the field from 1960 to 1970 have turned to other scientific interests or become managers. If history is a reliable leader, a resurgence should occur in another 20 years.

References

[1]E. C. Sullivan and W. C. Taylor, U.S. Pat. No. 36 136, 1915, and U.S. Pat. No. 1 304 623, 1919.
[2]W. Vogel, Struktur und Kristallisation der Gläser. VEB Deutscher Verlag für Grundstoffin-dustrie, Leipzig, German Democratic Republic, 1965; p. 78.
[3]W. E. S. Turner and F. Winks, *J. Soc. Glass Technol.*, **10**, 102 (1926).
[4]H. P. Hood and M. E. Nordberg, U.S. Pat. No. 2 106 744, 1934.
[5]T. H. Elmer, *Am. Ceram. Soc. Bull.*, **59**, 525 (1980).
[6]R. V. Lukas, U.S. Pat. No. 2 522 523, 1950.
[7]T. H. Elmer and M. E. Nordberg, Proceedings of the VII International Congress on Glass, Brussels, Belgium, June 28–July 3, 1965.
[8]M. Fanderlik, "Chemical Durability Defects of Borosilicate Glasses Caused by Improper Annealing Temperatures"; pp. 1–9 in International Commission on Glass and Ceramic Associ-ation of Japan, Symposium on Defects in Glass, Tokyo-Kyoto, Japan, September 1966.
[9]B. F. Howell, J. H. Simmons, and W. Haller, *Am. Ceram. Soc. Bull.*, **54**, 707 (1975).
[10]W. Skatulla, W. Vogel, and H. Wessel, *Silikat Technol.*, **9**, 51 (1958).
[11]R. Roy and C. Ruiz-Menacho; for abstract see *Am. Ceram. Soc. Bull.*, **38** [4] 229 (1959).
[12]R. Roy, *J. Am. Ceram. Soc.*, **430**, 670 (1960).

Opal Glasses

James E. Flannery and Dale R. Wexell

Corning Glass Works
Research & Development Div.
Corning, NY   14830

Opal glasses have a wide variety of commercial uses, with the most important being in food service, lighting, and cosmetics packaging. Opal glasses may be translucent or opaque. The degree of opacity depends on (1) the crystallite size, (2) the quantity of the opacifying phase, and (3) the refractive index difference between the precipitated phase and the glass matrix. Fluorine-containing glasses which contain NaF and/or CaF_2 phases are the most common opals, although some phosphate-containing opals have been commercially manufactured.

Opal glasses have a long and useful history, and some of the earliest glasses were opals. The ancient Egyptians produced opal glasses for jewelry and for many decorative effects. During the Middle Ages, opal glasses continued to be used decoratively, especially in cameos: The cameos were made by overlaying the opal on a dark glass and then grinding away the unneeded portion. Around the middle of the nineteenth century, opals began to be used on a large scale for lighting fixtures, and then for tableware.[1] White tableware and its "hospital-white" connotations particularly appealed to the American propensity for cleanliness. Today, opal glasses still have wide commercial application in food-service products and lighting fixtures.[2]

Opal glasses may be either translucent or opaque: The more densely opaque glasses are properly called alabaster glasses,[3] although this term is not widely used today. For the manufacturer, the major problem in making opal glasses is controlling the degree of opacity to yield a consistent, homogeneous product.

Types of Opal Glasses

Opal glasses are essentially phase-separated glasses, with their opacity resulting from light refraction and internal scattering between the separated phases.[4,5] This phase separation may be either liquid-liquid or liquid-crystal, with the latter more common commercially (cf. Figs. 1 and 2). Figure 1 is a photomicrograph of the cross section of an opal glass with a crystalline phase, and Fig. 2 represents a typical cross-sectional view of a liquid-liquid phase-separated opal. Here, it is apparent that the two liquids are continuous with each other, which also illustrates the durability problem common to liquid-liquid opals: Leach paths (as in attack by alkali) for the least-durable phase are continuous and permeate the structure. Excess degradation by alkaline solution usually occurs in these glasses.

A third type of opal glass may be formed by trapping small bubbles of gas within a glassy matrix; called bubble opals, these glasses have no commercial application at this time.

Selected compositions of commercial opal glasses are outlined in Table I. Note the diversity of opal types, complexity of compositions, and variations in physical properties.

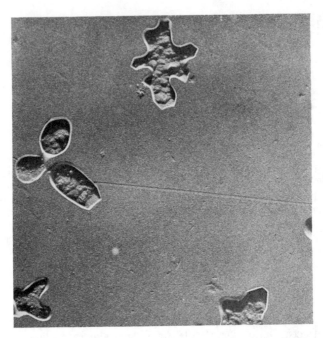

Fig. 1. Photomicrograph of liquid-crystal phase separation in opal glass.

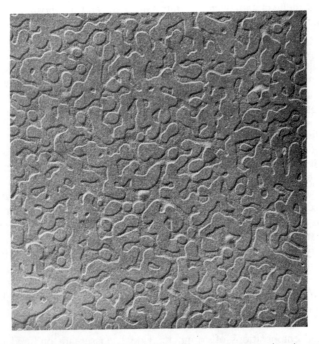

Fig. 2. Photomicrograph of liquid-liquid phase separation in opal glass.

Table I. Commercial Opal Compositions

Wt%	GEMCO*	DripCut†	Durand Table‡	Durand Arcopal§	BSN Rivanel¶	Sovirel Tamara**	Schott Inauer††	Code 6720‡‡	Rigopal§§
SiO_2	68.0	67.4	71.5	75.25	64.5	74.5	73.3	58.9	72.5
Al_2O_3	6.3	7.3	5.6	3.2	3.1	0.5	2.0	10.54	3.36
Na_2O	13.9	11.8	13.0	6.55	2.45	3.0	4.6	8.4	6.4
CaO	5.8	6.2		0.7	6.7		2.5	5.8	1.7
B_2O_3		0.6		13.45	12.6	12.75	14.5	1.35	13.7
F	5.2	5.2	3.7	0.8			0.65	4.2	0.5
P_2O_5							1.0		1.9
TiO_2					4.35	0.77	1.65		
MgO			6.3						
BaO									
ZnO						8.6		8.65	
ZrO_2					6.1				
S.P. (°C)	694	707	715	764	750	889	795	792	782
Exp. ($\times 10^{-7}$/°C, 25°–300°C)	91.1	86.6	75.2	43.6	44	29	38.2	79.5	46.5
Type of opal:	NaF	NaF	NaF	Liquid-liquid	Liquid-liquid	Liquid-liquid	Liquid-liquid (some phosphate)	CaF_2	Liquid-liquid (some phosphate)

*General Machine Co. of N.J., Middlesex, N.J.
†Jeanette Glass Co., Wheeling, W.V.
‡Durand Equipment and Mfg. Co., Durand, MI.
§Durand Equipment & Mfg. Co.
¶Rivanel Glass Co., France.
**Sovirel, Parret, France.
††Jenaer Glaswerk Schott, Jena, German Democratic Republic.
‡‡Corning Glass Works, Corning, NY.
§§Corning Glass Works.

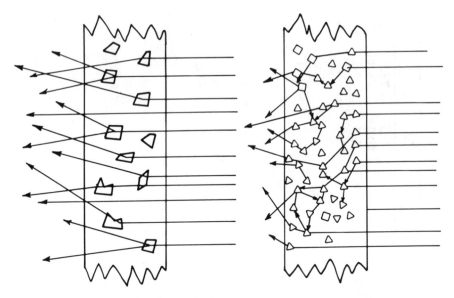

Fig. 3.   Light scattering in opal glass.

Crystalline opal glasses contain about 3–10 vol% crystallinity, whereas glass-ceramics, by definition, have 50 vol% or more crystallinity, and commonly over 90 vol%.[6] Crystallinity is usually estimated by determining crystalline cross-sectional area vs cross-sectional area of the glass phase in a cross-sectional photomicrograph.

The overall opacity of an opal glass is controlled by three factors[7]:

(1) The refractive-index difference between the two phases.

(2) The degree of phase separation (number of crystals in accord with principles of crystal growth (Tammann theory)[8,9] or volume of separated phase).

(3) The size and distribution of the separated phase.

Figure 3 illustrates the effect of these factors. Clearly, increasing the size and/or the number of crystallites or droplets of the separated phase increases internal scattering of light and, thus, opacity. Increasing the refractive-index difference between the two phases also increases internal scattering, thereby increasing opacity.

Opal glasses may also be classified as spontaneous or reheat opals. The spontaneous variety achieves opacity with initial cooling from the molten state and does not experience any substantial increased opacity with subsequent heat treatment. Reheat opals may be clear or translucent when cooled from the molten state; the final, denser opacity is obtained by further heat treatment. Figure 4 illustrates the nucleation and crystal-growth curves for both types of opals. The spontaneous opal-glass nucleation and crystal-growing curves overlap, allowing nucleation and crystal growth to occur simultaneously. The reheat opals must first be cooled through the crystal-growing range and held at or reheated to the nucleation temperature, then reheated to the crystal-growing temperature. For the manufacturer, the spontaneous opal offers considerable savings by eliminating the costly reheat cycle. It is, however, more difficult to achieve dense opacity (alabaster) with spontaneous opals.

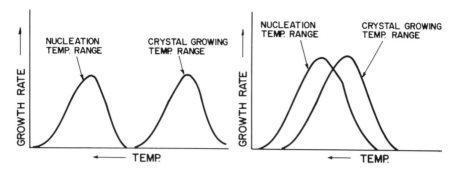

Fig. 4. Nucleation and crystal-growth curves for reheat and spontaneous opal glasses.

Most commercial opal glasses are spontaneous and crystalline. The most widely used crystalline opal glasses contain either calcium fluoride, sodium fluoride, or mixtures of the two species.[10-12] Generally, CaF_2 opals are spontaneous, whereas NaF opals are reheat. In contrast, no liquid-liquid opals are melted on a large scale. Crystalline or amorphous phosphate opals also have failed to reach commercial production in the United States.

The process of opalization is generally considered irreversible, i.e., the opacifying phase is not readily redissolved in the matrix glass until heated to above the softening point of the glass. However, Flannery[13] reported a system of alkali-metal-fluoride opal glasses that can be heated to transparency and cooled back to opacity without deformation of the glasses. Critical components for producing this phenomenon are MoO_3, WO_3, or As_2O_3.

Opacity and Opal Liquidus

For melting and forming opal glasses, one factor—the opal liquidus—is most critical. This factor may be defined as the cooling or reheating temperature at which opacity first appears. In the past, this temperature was estimated by human-eye detection of visible opacity in the molten glass. Sometimes spectrophotometric detection was used to increase resolution. It is obvious from the definition of opal liquidus that the cooling or heating rate causes shifts in it. A method has now been devised which maximizes the detection of subtle changes in glass opacity as a function of temperature and the cooling or heating rate.[14,15] Figure 5 shows a laser-reflectance apparatus for determining opal liquidus. Figure 6 shows a representative curve of a fluorophosphate opal with a *calcium apatite crystalline phase*[15] as the opacifying phase. A reading of 60 mV or more generally corresponds to what is "dense opacity" to the naked eye of an observer. The first high-temperature inflection point occurs around 1380°C (point A) and corresponds to the initial separation of the melted glass into two phases. This emulsification process results in increased scattering of the impinging laser beam. In the regions around B and C, crystallization occurs with the precipitation of various apatite phases, and two distinct apatite phases can be detected with high-temperature X-ray diffraction. A first-derivative curve can also be derived from this curve to yield a better value for each of the crystallization temperatures. The derivative curve will yield a sharp peak whenever there is a radical change in slope of the curve. The other curve represents a CaF_2 opal with no emulsification as a first step, but with a spontaneous precipitation of CaF_2 at 1010°C. Referring to the

145

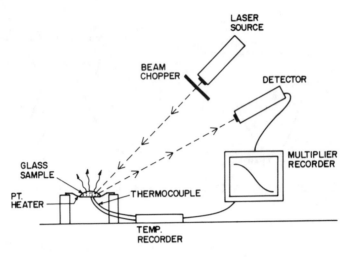

Fig. 5.   Laser-reflectance apparatus.

first curve, it is clear that if the opal liquidus is too high the glass will crystallize in the melting unit or during the forming process. This results in "brushmark" (elongated surface crystals) and often iridescence in the final product. If the opal liquidus is too low, little or no opacity will develop.

The opaqueness of opal glasses may be increased by adding known surface-tension modifiers, such as MoO_3 and As_2O_3, to the base-opal glass.[7] Amberg[16] earlier discussed the dramatic effect of molybdenum oxide in lowering the surface tension of silicate melts. Appen[17] later classified a series of oxides by its ability to

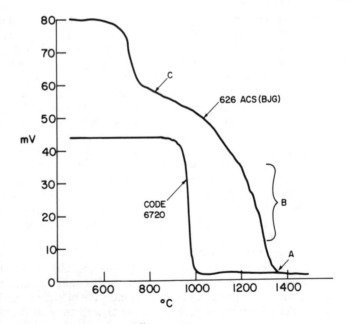

Fig. 6.   Laser-reflectance cooling curves.

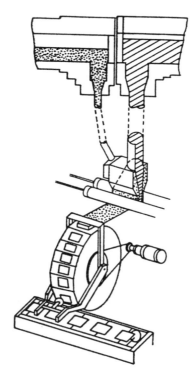

Fig. 7.   Corning process for forming laminated sheet.

lower the surface tension of silicate melts, and extended the classification to melts at 1300°C.[18] Surface-tension modifiers generally remain in the matrix phase, lowering its surface energy and helping the separated phase increase in size.

Forming Methods

All common forming methods[2] are used with opal glasses; pressing, sheet forming, and centrifugal casting produce the largest volumes. Blowing, which produces thin-walled articles, is used primarily for lighting globes. Figure 7 represents the Corning sheet-forming process,[19,20] which has produced more than one billion pieces of Corelle* opal dinnerware.

Weathering

Compositions for opal glasses may be complex, as shown earlier in Table I. For fluoride opals, the best opacity is usually developed at between 3 and 5% F. Below 3% opacity is weak, and above 5% the fluorine losses cause pollution problems and mold corrosion. With the desired opacity achieved, other factors become important: For example, for lighting and tableware products, weatherability becomes a concern. In many NaF opals, atmospheric moisture reacting with the glass can produce an oily, odorous film on the glass. This film absorbs gases and dust that destroy the esthetic value of the article. Figure 8 outlines the mechanism for this weathering phenomenon. The weathering can be attenuated by

---

*Corning Glass Works, Corning, NY.

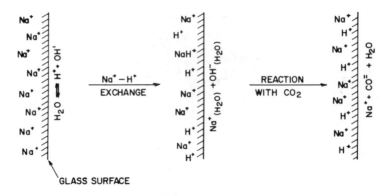

Fig. 8. Schematic of the weathering of NaF opals.

removing Na^+ from the glass surface[21] or applying a protective oxide (e.g., SnO_2) layer on the glass. For tableware, durability in the harsh environment of hot alkali, as in a dishwasher, presents a major problem. If the article is to be decorated with enamels, the softening point must be high enough to allow adequate curing of the enamels without distortion of the substrate. Some NaF opals also may actually fade or lose opacity during this enamel firing or during tempering.

"Flashmark"

Another major problem for opal-glass producers is "flashmark," which is more prevalent on spontaneous opals than on reheat opals. Flashmark is defined as a differential opacity, visible in reflected light, caused by uneven cooling during forming. Figure 9 illustrates this problem. Proper mold design for even cooling

Fig. 9. Flashmark in opal-glass production.

148

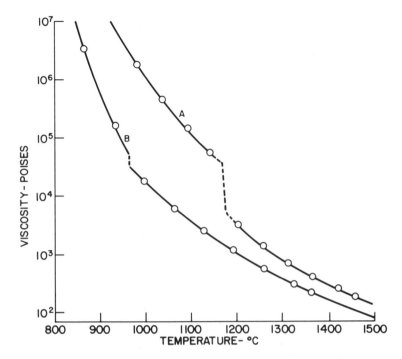

Fig. 10.   Viscosity vs temperature curves for crystalline opals.

does reduce the phenomenon. Adjusting the opal liquidus to the highest possible temperature compatible with melting and forming also alleviates this problem.

Viscosity

Figure 10 shows examples of opal-glass viscosity curves measured on a viscometer[†] with a cooling rate of 20°C/min. Curve A illustrates the sharp change in viscosity that may develop when crystals start to grow in a spontaneous opal; this can cause serious problems in forming. Curve B illustrates a more reasonable viscosity curve, and the crystal growth is not pronounced enough to cause forming problems.

Conclusion

Despite 4000 years of opal history, much is still unknown about the phase-separation phenomenon characteristic of these glasses. For example, we have not yet learned to nucleate selectively and control crystal growth, as is done in glass-ceramics. Fluoride-free commercial opals with dense opacity also have not been achieved.

The scientific and commercial futures of opal glasses remain "bright white."

References

[1]C. H. Commons, Jr., "Past and Present Practices and Theory of Opaque Glass," *Am. Ceram. Soc. Bull.*, **27**, 337–44 (1948).
[2]S. R. Scholes; pp. 76–77 in Modern Glass Practice, 7th rev. ed. Revised by Charles H. Greene. Cahners, Boston, 1975.

[†]Margules type, Corning Glass Works.

[3]K. Fuka and Y. Yoshioka, "Opal Glasses and Alabaster Glass and Their Applications," *J. Jpn. Ceram. Soc.*, **345**, 302–17 (1921).
[4]S. R. Scholes; pp. 326–29 in Modern Glass Practice. Cahners, Boston, 1975.
[5]P. F. James, "Nucleation in Glass Forming Systems — A Review"; pp. 1–48 in Advances in Ceramics, Vol. 4. Edited by J. H. Simmons, D. R. Uhlmann, and G. H. Beall. The American Ceramic Society, Columbus, OH, 1982.
[6]J. M. Stevels, "Glass Ceramics"; pp. 425–31 in Science of Ceramics, Vol. 2. Edited by G. H. Stewart. Academic Press, New York, 1965.
[7]J. E. Flannery, W. H. Dumbaugh, and G. B. Carrier, "Improving the Opacity of Phase Separated Opal Glasses by Alterations of the Interfacial Tension," *Am. Ceram. Soc. Bull.*, **54** [12] 1066–71 (1975).
[8]R. S. Bradley, "Nucleation in Phase Changes," *Q. Rev. (London)*, **5**, 315–43 (1951).
[9]D. Turnbull, "Phase Changes"; pp. 225–306 in Solid State Physics, Vol. 3. Edited by Frederick Seitz and David Turnbull. Academic Press, New York, 1956.
[10]G. Rothwell, "The Crystalline Phase in Fluoride Opal Glasses," *J. Am. Ceram. Soc.*, **39** [12] 407–14 (1956).
[11]R. J. Callow, "The Precipitation of Fluoride in Glassy Systems," *J. Soc. Glass Technol.*, **36**, 266–69 (1957).
[12]Q. A. Juma'a and J. M. Parker, "Crystal Growth in Fluoride Opal Glasses"; pp. 218–36 in Advances in Ceramics, Vol. 4. The American Ceramic Society, Columbus, OH, 1982.
[13]J. E. Flannery, "Spontaneous Fluoride Opal Glasses with Thermally Reversible Opacity," U.S. Pat. No. 3 667 973, June 6, 1972.
[14]J. L. Stempin and D. R. Wexell, "Laser Reflectance for Investigation of High Temperature Phase Separation," American Chemical Society Meeting, Washington, DC, August 29, 1983.
[15]J. E. Flannery, J. L. Stempin, and D. R. Wexell, "Opal Glasses Having an Apatite Opacifying Phase," U.S. Pat. No. 4 536 480, August 20, 1985.
[16]C. R. Amberg, "Effect of Molybdenum and Other Oxides on Surface Tension of Silicate Melts and on Properties of Refractories and Abrasives," *J. Am. Ceram. Soc.*, **29** [4] 87–93 (1946).
[17]A. A. Appen, "Attempt to Classify Components According to Their Effect on the Surface Tension of Silicate Melts," *Zh. Fiz. Khim.*, **26**, 1339–1404 (1952).
[18]A. A. Appen and S. S. Kayalowa; pp. 61–69 in Advances in Glass Technology, Part II. Edited by F. R. Matson and G. E. Rindone. Plenum, New York, 1963.
[19]J. W. Giffen, "Apparatus for Forming Articles from Overlapping Glass Sheets," U.S. Pat. No. 3 347 652, October 17, 1967.
[20]J. W. Giffen, D. A. Duke, W. H. Dumbaugh, Jr., J. E. Flannery, J. F. MacDowell, and J. E. Megles, "Glass Laminated Bodies Comprising a Tensilely Stressed Core and a Compressively Stressed Surface Layer Fused Thereto," U.S. Pat. No. 3 673 049, June 27, 1972.
[21]Kun-Er Lu and W. H. Tarcza, "Method for Improving the Durability of Spontaneous NaF Opal Glassware," U.S. Pat. No. 4 187 094, February 5, 1980.

Photosensitive Glasses

ROGER J. ARAUJO

Corning Glass Works
Research and Development Div.
Corning, NY 14831

Metal colloids can impart color to glass and nucleate crystallization. Anisotropic colloids can polarize light. Light can be used to suspend these colloids in any desired spatial pattern in a glass. Photosensitive glasses yielding multicolored images are described, as well as glasses containing silver-halide droplets, which respond reversibly to light.

Colored glasses can be obtained by the suspension of colloids, which absorb light. The famous Lycurgus cup, presently housed in the British Museum and believed to have been made by the Romans in the fourth century, owes its peculiar color to the suspension of silver and gold colloids. Dalton[1] found that the ruby color of certain glasses containing copper could be intensified by irradiating the glass with ultraviolet rays prior to a heat treatment used for developing color. Since that observation, photonucleation of metallic colloids and the interaction of metallic colloids with light have received a great deal of attention. These studies have resulted in the discovery of various new phenomena and the development of many new products.

It would be useful to review the way ruby glasses are formed: The glass is normally melted so that it contains ions of copper, silver, or gold in true solution,[2] as well as a certain level of thermal reducing agents such as arsenic, antimony, or tin. With certain heat treatments, the thermal reducing agent reduces some of the metal ions, and colored colloidal particles are formed.

Photography in Glass

The effect of irradiation on ruby-glass formation immediately suggested that these colloidal materials might be used photographically. Stookey[3] found that the efficacy of irradiation dramatically increased if cerium were included in the glass. He suggested that Ce^{3+} is photolyzed to produce Ce^{4+}: The electron is trapped by a noble metal ion, forming a neutral atom which acts as a nucleus for the thermal reduction of the metal necessary for colloid formation.

The advantages of using such a material as a photographic medium are fairly obvious: (1) The image formed is completely permanent; (2) because the colloids are small, the resolution capability of the material is extremely high; and (3) because the image has a finite thickness, interesting three-dimensional effects are sometimes observed.

Several interesting materials result from modifying photosensitive glasses. These derivatives are based on the observation that metal colloids can act as nucleating sites for other crystalline phases.[4] The precipitation of sodium fluoride has been nucleated to produce a white opal pattern for decorative purposes.

151

Photochemically Machinable Glasses

Under controlled conditions, colloidal silver can be used for nucleating lithium metasilicate. The latter phase is considerably more soluble in mineral acids than the glass from which it is formed. This new material allows preferential etching in a photographically produced pattern, thus affording a technique for very precise chemical machining of a glass.

Such glasses have been used for making screens containing more than a thousand precisely shaped and regularly positioned holes per square inch. Another application of chemical machining is in the production of fluidic devices depending on precisely shaped channels for their effectiveness. A flexographic printing technique uses chemically machined glasses as the master for producing a rubber printing plate. Using chemically machined glass in this way decreases the number of production steps and furthermore produces a higher-quality printing plate than those made by older techniques. More recently, chemically machined, photosensitive glass has been used for digital displays and magnetic recording heads.

Colored Glasses

The range of phenomena and products related to metallic colloids in glass has expanded dramatically because the optical properties of metal colloids depend on their shape, which need not be spherical. Stookey and Araujo[5] showed that when elongated silver particles were produced in glass the color varied with elongation, essentially as predicted by scattering theory. Specifically, the wavelength of maximum absorption moved very slowly to higher frequencies with increasing values of the elongation ratio when the electric vector was perpendicular to the long axis of the colloid, but moved rapidly toward lower frequencies for the parallel component (see Fig. 1). Hence, a glass containing oriented, elongated metallic particles acts as a polarizer over the wavelength interval in which the colloids absorb.

Polarization

Glasses containing metal colloids of uniform size and shape should exhibit very narrow absorption bands. Transmission bands can easily be produced in precisely the same spectral ranges by merely inserting a piece of this material between two ordinary polarizing sheets so that its long axis is at a 45 degree angle to each sheet and the two polarizing sheets are crossed. In such a configuration, optical rotation takes place only in the frequency range where absorption occurs in the polarizing glass; light is transmitted through the sandwich in this frequency range only. Thus, the narrow absorption band due to small particles in glass becomes the mechanism for producing a narrow transmission band. An important application of this phenomenon will be seen later.

If the glass contains particles with a continuous distribution of elongations, then it polarizes light throughout the visible region. Polarizing sunglasses are an attractive application of this phenomenon, since such glasses very effectively reduce glare. Light reflected from smooth plane surfaces can be highly polarized, and this reflected light is considered glare; the light scattered from less-regular shapes, such as a tree or a human body, is not highly polarized and ordinarily contains information useful to vision. Wearing polarizing sunglasses therefore removes glare and increases the signal-to-noise ratio of the visual system.

A third application of polarizing material, which has tantalized everyone for years, is in automobile windshields and headlights: At night the headlights of oncoming cars are attenuated, and a driver is saved from temporary blindness at the critical moments of passing. In any of these applications, glass has the advan-

152

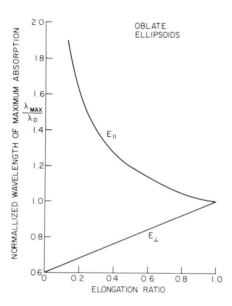

Fig. 1. Wavelength of maximum absorption vs elongation ratios of metal colloid.

tage of scratch resistance and chemical durability. In the last application, it also easily withstands the high temperatures associated with automobile headlights.

Photochromic Glasses

Further advantages of glasses containing elongated metal colloids are exhibited by a family of glasses discovered by Armistead and Stookey.[6] These glasses darken when irradiated with ultraviolet light, but differ from the other photosensitive glasses in resuming their original state after irradiation ends. This remarkable behavior arises from the photolysis of the minute particles of silver halide suspended in the glass matrix. As in photolysis of photographic emulsions, irradiating these glasses results in the formation of tiny anisotropic silver specks on the surface of the silver-halide particles. Araujo[7] explained that the photolysis of these glasses differs from that of photographic emulsions in that hole trapping involves copper ions, thereby making the photoreactions reversible.

Photochromic glasses are commercially applied as automatic sunglasses. Araujo et al.[8] demonstrated that stretching the glass to produce elongated silver-halide particles makes glasses that are both photochromic and, in their darkened state, polarizing. Unfortunately, this fabrication technique does not easily adapt to mass production and these glasses, although possibly constituting the ideal automatic sunglass material, have not been introduced to the marketplace.

Stretching is not the only way of making a glass that is polarizing in its darkened state. As already noted, anisotropic specks of silver form when photochromic glasses are darkened by irradiation. Polarization is not ordinarily observed, however, because the elongated specks are randomly oriented in space. Incomplete optical bleaching with polarized light results in a distribution of orientations that is no longer random[9] and, therefore, produces a polarizing glass. For reasons not yet completely understood, some glasses so treated exhibit a peculiar memory effect. After having been completely faded in the dark, these glasses,

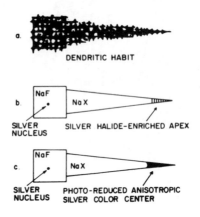

a. DENDRITIC HABIT

b. NaF · SILVER NUCLEUS   Na X   SILVER HALIDE-ENRICHED APEX

c. NaF · SILVER NUCLEUS   Na X   PHOTO-REDUCED ANISOTROPIC SILVER COLOR CENTER

Fig. 2. Schematic illustration of photocrystallites.

when redarkened by unpolarized light sources such as the sun, again become polarizers. Hence, for certain glasses, optical bleaching with intense polarized light is a simple way of producing photochromic glasses which, in their darkened state, are also polarizing.

Unfortunately, the memory exhibited by these glasses is imperfect, and the polarization efficiency decreases slightly with each darkening and bleaching cycle, until after about 100 cycles useful polarizing efficiencies are no longer exhibited. Of course, darkening and fading rates are completely unaffected by this gradual change in polarizing efficiency.

Since the wavelength of light absorbed by an anisotropic metal speck varies with its elongation, one would expect to be able to alter the color of a previously darkened photochromic glass by incomplete bleaching with irradiation in narrow wavelength intervals. Such is actually the case, and multicolored pictures have been recorded in photochromic glasses having extremely slow thermal bleaching or dark fading rates.

Araujo et al.[10] demonstrated that it is possible, by chemical reduction, to obtain anisotropic specks of silver on the surface of small particles of silver halide suspended in glass without ultraviolet irradiation. Glasses containing such particles can be optically bleached to produce the same colors or polarizing effects as can photochromic glasses in their darkened state. In the chemically reduced glasses, of course, the colors and polarization are permanent as long as the glass is kept out of very bright light.

Recent Advances

Optical bleaching with polarized light of either darkened, slow-fading photochromic glasses or thermally reduced glasses provides an attractive way of storing information which can be displayed with a very high signal-to-noise ratio. As discussed earlier, such information is in the form of light-transmitting spots in a very dark background, when viewed between cross polaroids.

No discussion of photosensitive glasses or glasses containing anisotropic metal specks would be complete without mentioning the polychromatic glasses invented by Stookey and Pierson.[11] A series of ultraviolet exposures and heat treatments can form multicolor photographic patterns. Stookey et al.[12] elucidated the mechanism of color formation: Pyramidal crystals of sodium fluoride are formed, and the sharp tips are decorated by a coating of colloidal silver (see Fig. 2). The sharpness of the tip determines the effective elongation ratio of the silver colloid and therefore, as explained previously, the spectral region absorbed.

References

[1] R. H. Dalton, U.S. Pat. No. 2 326 012, 1945.
[2] S. D. Stookey, *J. Am. Ceram. Soc.*, **32**, 246 (1949).
[3] S. D. Stookey, *Ind. Eng. Chem.*, **41**, 856 (1949).
[4] S. D. Stookey, *Ind. Eng. Chem.*, **45**, 115 (1953).
[5] S. D. Stookey and R. J. Araujo, *Appl. Opt.*, **7**, 777 (1968).
[6] W. H. Armistead and S. D. Stookey, *Science*, **144**, 150 (1964).
[7] R. J. Araujo, *Contemp. Phys.*, **21**, 77 (1980).
[8] R. J. Araujo, W. H. Cramer, and S. D. Stookey, U.S. Pat. No. 3 540 793, 1970.
[9] N. F. Borrelli, J. B. Chodak, and G. B. Hares, *J. Appl. Phys.*, **50**, 5978 (1979).
[10] R. J. Araujo, N. F. Borrelli, J. B. Chodak, G. B. Hares, G. S. Meiling, and T. P. Seward, U.S. Pat. No. 4 125 405, 1978.
[11] S. D. Stookey and J. E. Pierson, U.S. Pat. No. 4 017 318, 1977.
[12] S. D. Stookey, G. H. Beall, and J. E. Pierson, *J. Appl. Phys.*, **49**, 5114 (1978).

Glass-Ceramics

GEORGE H. BEALL

Corning Glass Works
Glass Ceramic Research Dept.
Corning, NY   14830

The composition, structure, properties, and applications of the key glass-ceramic types in commercial use: β-spodumene, β-quartz, cordierite, nepheline, lithium silicate, and fluormica are reviewed. The basic steps in the manufacturing process: melting, forming, crystallization, and finishing, are described for each case. Examples of restrictions and compromises imposed by manufacturing on materials properties are given.

Glass-ceramics are microcrystalline solids produced by the controlled devitrification of glass. Glasses are melted, fabricated to shape, and then converted by heat treatment to a predominantly crystalline ceramic. The basis of controlled crystallization lies in efficient internal nucleation,[1] which allows development of fine, randomly oriented grains without voids, microcracks, or other porosity.

A unique manufacturing advantage of glass-ceramics over conventional ceramics is the ability to use high-speed plastic forming processes developed in the glass industry (e.g., pressing, blowing, rolling, etc.) to create complex shapes essentially free of internal inhomogeneities. Because glass-ceramic compositions are designed to crystallize, however, they cannot be held at temperatures below the liquidus during the forming process. Therefore, the viscosity at the liquidus temperature is critical both in the choice of a forming process and in the choice of a glass composition.

There are currently six basic composition systems from which commercial glass-ceramics of economic importance are made: (1) Li_2O-Al_2O_3-SiO_2, glass-ceramics of very low thermal expansion coefficient; (2) MgO-Al_2O_3-SiO_2, cordierite glass-ceramics of good mechanical, thermal, and dielectric properties; (3) Li_2O-SiO_2 glass-ceramics with photochemical etching capability; (4) Na_2O-Al_2O_3-SiO_2, nepheline glass-ceramics with high mechanical strength from compression glazing; (5) K_2O-MgO-Al_2O_3-SiO_2-F, machinable fluormica glass-ceramics; and (6) CaO-MgO-Al_2O_3-SiO_2, inexpensive glass-ceramics from natural materials and slags.

Glass-Ceramics Based on Li_2O-Al_2O_3-SiO_2

The base system Li_2O-Al_2O_3-SiO_2 has produced glass-ceramics of the lowest thermal expansion coefficient based on either β-quartz or β-spodumene (keatite) solid solution crystal phases. The β-quartz solid solutions are metastable crystalline phases of the general composition $(Li_2,R)O \cdot Al_2O_3 \cdot nSiO_2$,[2] where n varies from 2 to 10 and R is a divalent cation, normally Mg^{2+} or Zn^{2+}. Near the stoichiometric end-member composition, $LiAlSiO_4$, this crystal phase is stable and is referred to as β-eucryptite. In commercial compositions, however, n is always in the range of 4 to 7; the β-quartz structure is metastable, and will break down to other phases if heated above 900°C. Substitutions of MgO and ZnO for Li_2O and $AlPO_4$ for SiO_2 are useful in reducing both batch cost and liquidus temperature, the latter generally improving glass stability.

The combination of TiO_2 and ZrO_2 has been shown most effective in nucleation of β-quartz from lithium aluminosilicate glasses,[3] and is used in most commercial β-quartz glass-ceramics. When used at levels near 2 mol% each, very fine crystals ($\leq$100 nm) are achieved and a transparent, highly crystalline body can be produced. Both fine crystal size and low birefringence inherent in β-quartz solid solutions allow light scattering to be minimized.[4] Stuffed derivatives of β-quartz, where tetrahedral Al^{3+} replaces Si^{4+} in the network and Li^+ ions support the high quartz structure, are characterized by very low thermal expansion coefficients, generally near zero or sometimes even negative. The combination of transparency, very low thermal expansion behavior, optical polishability, and strength greater than glass has generated applications such as cookware, telescope mirror blanks, woodstove windows, and infrared-transmitting range tops.

Table I lists the compositions of three commercial glass-ceramics of the β-quartz solid solution type from different manufacturers along with their areas of application. The ingredients that compose the crystal phase are separated from those that concentrate in the residual glass, which makes up less than 10 vol% of the body. The latter, generally tramp constituents, like soda and potash, form a persistant aluminosilicate glassy phase along grain boundaries. Arsenic oxide is added as a refining agent to purge gas bubbles and various colorant transition metal oxides are present and provide, in concert with titania, a brownish tint.

Table I. Composition of Transparent Glass-Ceramics Based on β-Quartz Solid Solution (Wt%)

		VISION Corning	ZERODUR* Schott	Narumi* Nippon Electric
SiO_2	x1	68.8	55.5	65.1
Al_2O_3		19.2	25.3	22.6
Li_2O		2.7	3.7	4.2
MgO		1.8	1.0	0.5
ZnO		1.0	1.4	
P_2O_5			7.9	1.2
F				0.1
Na_2O	gl	0.2	0.5	0.6
K_2O		0.1		0.3
BaO		0.8		
TiO_2	n	2.7	2.3	2.0
ZrO_2		1.8	1.9	2.3
As_2O_3	f	0.8	0.5	1.1
Fe_2O_3	c	0.1	0.03	0.03
CoO		50 ppm		
Cr_2O_3		50 ppm		
		Transparent cookware	Telescope mirrors	Rangetops Stove windows

*As analyzed at Corning Glass Works, x1, oxides concentrated in crystal; gl, oxides concentrated in glass, n, nucleating-agent oxides, f, fixing-agent oxide; c, colorant oxides.

Figures 1 and 2 show the appearance and microstructure of the VISIONS* cookware product. This material is highly transparent because of the ultrafine grain size ($\approx$60 nm ($\approx$600 Å)) and low birefringence of the constituent β-quartz solid solution crystals (Fig. 2). This transmission electron microphotograph illustrates the highly crystalline microstructure with a high particle density of $5 \times 10^{21}/m^3$. The nucleating phase, $ZrTiO_4$, is visible as small specs scattered throughout. This glass-ceramic has a coefficient of thermal expansion of $7 \times 10^{-7}/°C$ ($0°–500°C$), thus giving the product outstanding thermal shock resistance.

Opaque glass-ceramics are readily achieved in the system Li_2O-Al_2O_3-SiO_2 by crystallizing heterogeneous-nucleated glasses at relatively high temperatures ($1000°–1200°C$) and allowing development of the stable crystalline assemblage, which always includes β-spodumene solid solution as the main phase.[5] This tetragonal crystal, like its hexagonal predecessor, has very low thermal expansion coefficient. It is a stuffed derivative of the silica polymorph keatite, and is sometimes referred to as stuffed keatite or keatite solid solution. Its composition ranges from $Li_2O \cdot Al_2O_3 \cdot 4SiO_2$ to $Li_2O \cdot Al_2O_3 \cdot 10SiO_2$. Considerable substitution of magnesium for lithium ($Mg^{2+} \rightleftharpoons 2Li^+$) is permitted in the structure, though less than is allowed in the metastable β-quartz precursor.[2] The transformation from β-quartz to β-spodumene solid solution usually occurs between 900° and 1000°C, is irreversible, and is accompanied by an increase in grain size (usually five- to tenfold). When TiO_2 is used as a nucleating agent, rutile development accompanies

*Corning Glass Works, Corning, NY.

Fig. 1. VISIONS cookware, a transparent β-quartz solid solution glass-ceramic product.

159

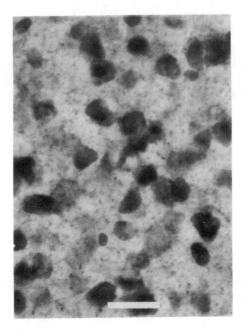

Fig. 2. Microstructure of VISIONS as revealed by TEM (bar = 100 nm).

Fig. 3. Microstructure of CORNINGWARE cookware, an opaque β-spodumene-rutile glass-ceramic.

the silicate phase transformation. Because of the high refractive index and bi-refringence of this phase, a high degree of opacity is developed.

Typically, β-spodumene glass-ceramic grains are in the 1–2 μm range (Fig. 3). Secondary grain growth is sluggish and generally linear with the cube root of time.[6] The resistance to grain growth is particularly important in view of the marked anisotropy in thermal expansion of β-spodumene crystals (Fig. 4).

Table II lists the compositions of two commercial β-spodumene glass-ceramics, one used for CORNINGWARE[†] cookware, and one for ceramic regenerators in turbine engines. The former shows a multicomponent glass containing titania as the basic nucleation agent. Magnesia partially substitutes for lithia to lower the liquidus temperature, thereby allowing sufficient viscosity for pressing, blowing, and tube drawing. Figure 5 illustrates the viscosity temperature curve (°C). Since the liquidus temperature is about 1230°C, the viscosity at the liquidus can be seen to be near 10^5 P, sufficient for most glass-forming processes.

This commercial glass-ceramic is crystallized with a maximum temperature near 1125°C, about 100°C below the liquidus. It is highly crystalline (>93%) and contains β-spodumene as the major phase with minor spinel, rutile, and glass. The thermal expansion coefficient is $12 \times 10^{-7}/°C$ (0°–500°C), and the abraded flexural strength about 100 MPa.

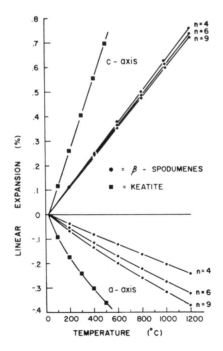

Fig. 4. Thermal expansion behavior of β-spodumene solid solution crystals ($Li_2O \cdot Al_2O_3 \cdot nSiO_2$) along tetragonal axes (after Ostertag et al. (Ref. 7)).

---

†Corning Glass Works.

Table II. Composition of Glass-Ceramics Based on β-Spodumene Solid Solution

		Corningware		CERCOR Corning	
		(wt%)	(mol%)	(wt%)	(mol%)
SiO_2	⎫	69.7	73.6	72.5	75.9
Al_2O_3	⎪	17.8	11.0	22.5	13.6
Li_2O	⎬ x1	2.8	5.9	5.0	10.5
MgO	⎪	2.6	4.1		
ZnO	⎭	1.0	0.8		
Na_2O	⎫ gl	0.4	0.4		
K_2O	⎭	0.2	0.1		
TiO_2	⎫ n	4.7	3.7		
ZrO_2	⎭	0.1	0.1		
Fe_2O_3	c	0.1	0.1		
As_2O_3	f	0.6	0.2		
		Cookware Hot plates		Heat exchangers Regenerators	

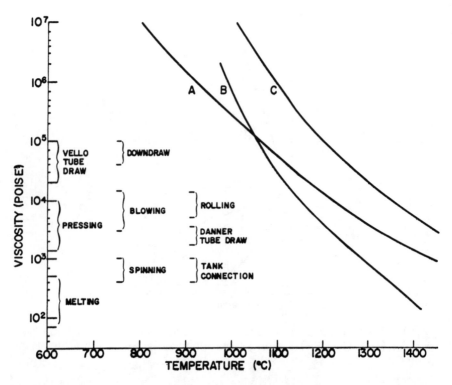

Fig. 5. Viscosity-temperature curves of some crystallizable glasses and viscosity ranges for glass-forming processes.

The regenerator glass-ceramic is produced as a honeycomb product for a turbine engine heat exchanger. It is made from a powdered glass frit, which as a slurry can impregnate paper wound as alternately corrugated sheets on a wheel.[8] After firing, a porous ceramic regenerator wheel is produced which allows energy to be transferred from hot exhaust gases to the cold intake air in a turbine engine. The very low thermal expansion ($\approx 5 \times 10^{-7}/°C$ (0°–1000°C), and high thermal stability (>1200°C) are important in this application. Since very low thermal expansion and the highest thermal stability are required, no extra components are permitted in the original glass composition.

MgO-Al$_2$O$_3$-SiO$_2$ Glass-Ceramics Based on Cordierite

Glass-ceramics based on the hexagonal form of cordierite, sometimes referred to as indialite, are strong, have excellent dielectric properties, and good thermal stability and shock resistance. Corning Code 9606, whose composition is given in Table III, is the standard glass-ceramic used for radomes. It is a multiphase material nucleated with titania, but based on cordierite of composition $Mg_2Al_4Si_5O_{18}$ with some solid solution toward "Mg-beryl" (i.e., $Mg^{2+} + Si^{4+} \rightleftharpoons 2Al^{3+}$).[9] This major phase is mixed with cristobalite, rutile, magnesium dititanate, and minor glass, which is isolated at grain-boundary nodes. The mechanical properties of these glass-ceramics have been studied extensively. A Weibull plot of flexural strength data on transverse-ground bars hewn from a slab of this commercial composition is shown in Fig. 6. Other important properties include a coefficient of thermal expansion (0°–700°C) of $45 \times 10^{-7}/°C$, fracture toughness (K_{IC}) 2.2 MPa m$^{1/2}$, thermal conductively 0.009 cal/s·cm·°C, Knoop hardness 700, dielectric constant and loss tangent at 8.6 GHz: 5.5, and 0.0003, respectively.

One of the difficulties in crystallization of glass-ceramics involves relieving stresses due to change in density accompanying phase transformation. This is well illustrated by Code 9606. Table IV shows the phase assemblage and corresponding density when the parent glass is heated to various temperatures for two hours. There is a significant increase in density from glass to dense metastable crystalline assemblage up to 1010°C, followed by a volume expansion to cordierite above this temperature. Clearly, to avoid extreme stresses and cracking, the heat-treatment schedule must be carefully adjusted to minimize extremes in metastable phase

Table III. Commercial Cordierite Glass-Ceramic (Corning 9606)

Composition		wt%	mol%	Phases
SiO$_2$		56.1	58.1	
Al$_2$O$_3$	x1	19.8	12.1	Cordierite
MgO		14.7	22.6	Cristobalite
				Rutile
CaO		0.1	0.1	Mg-dititanate
TiO$_2$	n	8.9	6.9	
As$_2$O$_3$	f	0.3	0.1	
Fe$_2$O$_3$		0.1	0.1	

Use: Radomes

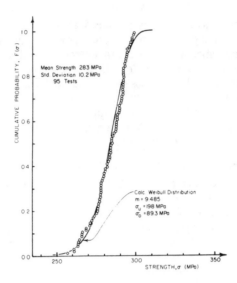

Fig. 6. Flexural strength distribution of transverse ground bars of Corning Code 9606 glass-ceramic (after Lewis et al. (Ref. 10)).

Table IV. Phase Assemblages during Crystallization of Glass-Ceramic 9606

Temp. (°C)	Density	Phases
700	2.64	Glass
800	2.67	Glass, $MgTi_2O_5$
900	2.75	β-quartz ss, $MgTi_2O_5$, glass
1010	2.95	α-quartz, sapphirine, enstatite, $MgTi_2O_5$, rutile
1260	2.60	Cordierite ss, rutile, $MgTi_2O_5$

density and allow sufficient plastic glassy phase at various stages to prevent cracking. The final desired assemblage developed at 1260°C has good thermal stability toward grain growth and will not revert to other phases when held at lower temperatures.

The choice of composition for Code 9606 was based primarily on glass-forming considerations. To optimize viscosity at the liquidus, the lowest ternary eutectic in the refractory system $MgO-Al_2O_3-SiO_2$ was approached with little compromise in the key properties of cordierite by maintaining it as the major crystalline phase. Some cristobalite had to be incorporated, which had the adverse effect of raising thermal expansion. This phase, however, allowed a post-ceram surface leaching treatment with hot caustic to be effective in producing a porous silica-deficient skin which tends to prevent mechanical flaw initiation.[10]

The viscosity-temperature curve for the parent glass of Code 9606 appears as curve A in Fig. 5. Because the liquidus temperature is near 1350°C, close to the ternary eutectic temperature, and the glass is relatively low in silica (58 mol%) and therefore fluid, only such forming processes as spinning or other types of casting can be used. Fortunately, the radome shape is particularly amenable to centrifugal casting.

Photosensitive Lithium Silicate Glass-Ceramics

Homogeneous nucleation of metal particles in glass can be catalyzed by photoelectrons provided by the action of ultraviolet light on special compositions containing small amounts of gold and silver ions.[11] Cerous ions are generally added as an "optical sensitizer" because they absorb ultraviolet light and donate photoelectrons to produce metal atoms according to:

$$(Au, Ag)^+ + Ce^{3+} \xrightarrow{uv} Ce^{4+} + (Au, Ag)^0 \tag{1}$$

In certain lithium silicate glasses containing such ions, areas exposed to ultraviolet light develop metal colloids on heat treatment, which then selectively nucleate a dendritic form of the crystal lithium metasilicate (Li_2SiO_3). This metastable crystal is far more easily etched in hydrofluoric acid than is the parent glass, allowing an irradiated pattern to be selectively etched out.[12] The resulting photoetched glass can then be flood-exposed to ultraviolet rays and heat-treated beyond that temperature region of the metastable lithium metasilicate phase; the stable phase, lithium disilicate, is then produced. The resulting glass-ceramic is strong, tough, and faithfully replicates the original photo-etched pattern.

The commercial photo-etchable glass-ceramic is Corning Code 8603, whose composition is given in Table V. This composition has far less lithia (9.3 wt%) than that required to produce a completely crystalline lithium disilicate glass-ceramic ($\approx$20 wt%). The lithia level has been lowered to minimize cost to about 9%. This is about the minimum necessary to allow development of a continuous dendritic pattern of lithium metasilicate after exposure, nucleation for an hour at 510°, and heat treatment at 610°C for two hours. Figure 7 illustrates the etchable microstructure in this commercial material, which is referred to as Fotoform.[‡] The

Table V. Commercial Photosensitive Glass-Ceramics

		Fotoform/Fotoceram	
		Corning 8603	
		wt%	
SiO_2		79.6	
Al_2O_3		4.0	
Li_2O		9.3	
K_2O		4.1	
Na_2O		1.6	
Ag	n	0.11	Nucleant
Au		0.001	Prenucleant
CeO_2	s	0.014	Optical sensitizer
SnO_2		0.003	Thermal sensitizer
Sb_2O_3	f	0.4	Thermal sensitizer and fining agent

---

‡Corning Glass Works.

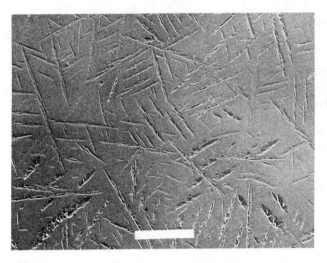

Fig. 7. Microstructure of Fotoform glass-ceramic as revealed by REM (bar = 1 μm).

role of potash and alumina is to stabilize the glass with respect to increasing viscosity at the liquidus. Tin oxide is useful as a thermal sensitizer which enhances metal colloid development, thus improving nucleation efficiency and ultimate resolution. Antimony oxide is effective as a refining agent and also acts in concert with tin oxide as a thermal sensitizer.

After the Fotoform stage, the etched glass is exposed and reheated beyond the lithium metasilicate stage to 850°C for two hours to develop the stable disilicate phase. At this point the final glass-ceramic is composed of $Li_2Si_2O_5$, α-quartz, with considerable potassium aluminosilicate residual glass remaining. This material has good abraded flexural strength ($\approx$150 MPa), and thermal expansion in the range of 120×10^{-7}/°C, matching ferrites and some metals. It is referred to as Fotoceram.§ Because the lithium ions are immobilized in the disilicate phase, the dielectric properties are good. In fact the high-temperature resistivity values of Fotoceram are more than two orders of magnitude higher than those of the parent glass.[13]

Numerous applications of Fotoform and Fotoceram have been developed, some of which require high resolution (hole or line spacing <0.01 mm). They include magnetic recording head pads, fluidic devices, cellular faceplates for gas-discharge displays, and charge plates for ink-jet printing.

Glazed Nepheline Glass-Ceramics

Fine-grained glass-ceramics based on soda nepheline ($NaAlSiO_4$), the end-member in the natural solid solution series which contains considerable K $\rightleftharpoons$ Na solid solution, have been formed from glasses in the simple system SiO_2-Al_2O_3-Na_2O-TiO_2.[14] Titania acts as an internal nucleating agent and opacifier, and the resulting phase assemblage typically contains nepheline, anatase, and some residual glass. These glass-ceramics are high in thermal expansion coefficient,

<hr>

§Corning Glass Works.

Table VI.   Nepheline Glass-Ceramic Tableware (Pyroceram 9609)

	Composition		Phases
	wt%	mol%	
SiO_2	43.3	53.0	
Al_2O_3	29.8	21.4	Nepheline, $NaAlSiO_4$
Na_2O	14.0	16.0	Celsian, $BaAl_2Si_2O_8$
BaO	5.5	2.6	Anatase TiO_2
TiO_2	6.5	6.0	Glass
As_2O_3	0.9	0.4	

reflecting that property of the major nepheline phase, which is structurally related to the high-expansion silica polymorph tridymite.[15]

Pyroceram[†] brand tableware is the commercial glass-ceramic based on nepheline; its body composition is given in Table VI. This material contains barium and arsenic oxides in addition to the nucleating agent titania. The role of the arsenic is in refining the glass, but the use of baria is more complex. It promotes a secondary aluminosilicate phase, celsian ($BaAl_2Si_2O_8$), which has a lower thermal expansion than nepheline, improving the thermal shock resistance of the glass-ceramic. It also lowers the liquidus temperature, allowing sufficient viscosity during the glass-forming stage to press cups and plates.

A key requirement for quality tableware, however, is a high-gloss surface, such as that produced by a glaze. To apply a glaze to a glass-ceramic, two conditions must be met: First, the glaze must be able to mature at temperatures below which the body will deform and, second, the glaze must be of lower thermal expansion than the body in order to preserve or enhance the strength of the articles.[16] With Code 9609 glass-ceramic, which has a thermal expansion coefficient of $95 \times 10^{-7}/°C$ (0°–300°C) and an upper crystallization temperature of 1100°C, a durable lead-lime-alkali aluminoborosilicate glaze some 30 points lower in thermal expansion ($\approx 65 \times 10^{-7}/°C$) can be applied and matured well below the point of deformation of the body. Because of this expansion difference between body and glaze below the strain point of the glaze ($\approx 500°C$), compression is developed at the surface, which increases the flexural strength of the glazed articles by about a factor of two. Figure 8 illustrates the relaxation expansion curves of glaze and body, which result in a working strength of 240 MPa in the tableware. Figure 9 shows the microstructure of the body, interface, and glaze, which is applied as a uniform coating, about 0.1 mm thick. A glaze-body reaction zone, some 10 μm thick, composed of crystals of plagioclase ($NaAlSi_3O_8$-$CaAl_2Si_2O_8$ solid solution) is evident. The interface crystals are elongated and several times larger than those of the body, but this apparently has no adverse effect on strength.

The major area of application for Pyroceram tableware is currently in the institutional segment (hotels, restaurants, etc.) where toughness and durability are critical issues.

Glass-Ceramics Based on Fluormica

Machinable glass-ceramics based on internally nucleated fluormica crystals in glass were originally described in the early 1970s.[17,18] One commercial material,

---

[†]Corning Glass Works.

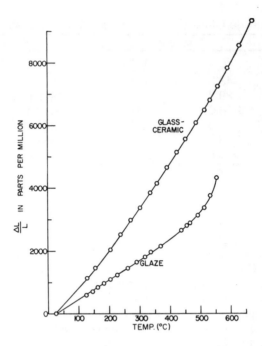

Fig. 8. Thermal expansion curves of nepheline glass-ceramic body and glaze (Corning Code 9609) (after Ref. 16)).

Macor,** has been marketed for over ten years, and has found wide application in such diverse and specialty areas as precision electrical insulators, vacuum feedthroughs, windows for microwave tube parts, sample holders for field ion microscopes, seismograph bobbins, gamma ray telescope frames, and boundary retainers on the space shuttle. The precision machinability of Macor with conventional, metal-working tools is directly due to the fine-grained house-of-cards microstructure (Fig. 10), where randomly oriented and flexible flakes tend to either arrest fractures or cause deflection or branching of cracks. Therefore, only local damage results, as tiny polyhedra of glass are dislodged. High dielectric strength (≈ 40 kV/nm) and very low helium permeation rates are also important in high-vacuum applications.

While the Macor glass-ceramic is based on the fluorine-phlogopite phase ($KMg_3AlSi_3O_{10}F_2$), recently another commercial material, Dicor,†† developed for use as dental restorations, has been released. This material, with improved chemical durability and translucency, is based on the tetrasilicic mica $KMg_{2.5}Si_4O_{10}F_2$. Table VII gives the composition ranges of these commercial glass-ceramics. The Macor composition is melted to an opal glass which is subsequently crystallized by heating to a top temperature of 950°C. It goes through a metastable crystallization sequence starting with the phase chondrodite, $2Mg_2SiO_4 \cdot MgF_2$, which crystallizes in the magnesium-rich matrix at the interface of the aluminosilicate droplets which constitute the parent opal glass.[19] The chondrodite subsequently

---

** Corning Glass Works.
†† Corning Glass Works.

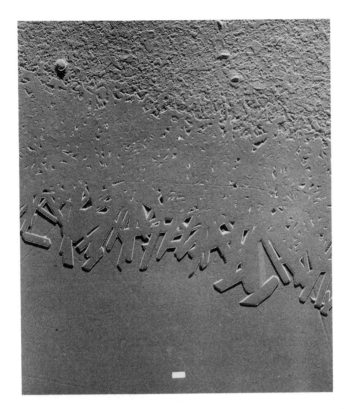

Fig. 9. Microphotograph (REM) showing nepheline glass-ceramic body, glazed-body interface with elongated plagioclase crystals, and non-crystalline glaze (Corning Code 9609) (bar = 1 μm).

transforms to norbergite, $MgSiO_4 \cdot MgF_2$, which finally reacts with the components in the residual glass to produce the fluorphlogopite mica and minor mullite. The role of the B_2O_3 in the composition is to enhance nucleation through stimulation of the amorphous phase separation, and to promote anisotropic crystal growth by lowering the viscosity of the glassy growth medium.

The Dicor product, on the other hand, is transparent in its parent-glass state. When heat-treated the mica is nucleated directly,[20] perhaps as a result of an extremely fine amorphous phase separation. The development of good strength ($\approx$150 MPa) is associated with the development of anisotropic flakes at relatively high temperature ($>$1000°C). Translucency is achieved by roughly matching crystal and glass indices and maintaining a fine-grained ($\approx$1 μm) crystal size. Ceria is added to simulate the fluorescent character of natural teeth. The unique features of Dicor for dental restorations include the close match to natural teeth in both hardness and appearance (Fig. 11). The glass-ceramic may be accurately cast using a lost-wax technique and conventional dental laboratory investment molds. The material's high strength and low thermal conductivity make it advantageous over conventional metal-ceramic systems.

The melting of high fluorine glasses, such as those producing mica glass-ceramics, generally requires electrical or cold-crown melting to limit vola-

169

Fig. 10. Microstructure of Macor machinable glass-ceramic (bar = 1 μm).

Table VII. Commercial Fluormica Glass-Ceramic Compositions

		Macor (Corning)	Dicor (Dentsply)
		wt%	wt%
SiO_2		47.2	56–64
B_2O_3	gl	8.5	
Al_2O_3		16.7	0–2
MgO		14.5	15–20
K_2O		9.5	12–18
F		6.3	4–9
ZrO_2	gl		0–5
CeO_2			0.05
Mica type:			
Macor		$K_{1-x}Mg_3Al_{1-x}Si_{3+x}O_{10}F_2$	
Dicor		$K_{1-x}Mg_{2.5+x/2}Si_4O_{10}F_2$	
		$x < 0.2$	

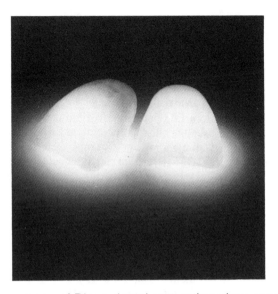

Fig. 11. Appearance of Dicor, dental restoration glass-ceramic.

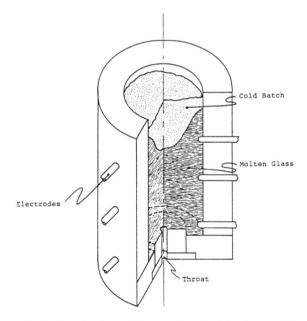

Fig. 12. Vertical electrical melting unit suitable for melting fluoride-containing glasses under conditions of low volatilization.

tilization. Such a furnace is illustrated in Fig. 12. The crystallization cycle may also cause enough fluorine loss from the surface to create a hard skin deficient in the mica phase. This thin skin can be removed by chemical etching or physical abrasion.

Inexpensive Glass-Ceramics Based on Slags

For two decades, inexpensive glass-ceramics based on blast furnace slags have been produced in Eastern Europe, particularly in the U.S.S.R. and in Hungary. These materials are usually rolled as sheet or cast as tiles, and are used for both interior and exterior wall cladding and flooring. Referred to as "slag-sitall" in the Soviet Union, these glass-ceramics presently constitute the largest volume application for crystallized glass.

Locsei[21] originally studied the system $Na_2O-CaO-MgO-Al_2O_3-SiO_2$ using various slags with additions of sulfides of heavy metals as nucleating agents. He synthesized a useful glass-ceramic called "Minelbite" characterized by high wear resistance and good chemical durability. An improved product, "Minelbite 2," contains substantial raw feldspar in the batch in addition to slag. The chemical analysis of this material is given in Table VIII.

Slag glass-ceramics are melted near 1450°C, formed into a glass, and subsequently heat-treated to a maximum temperature near 1000°C. A nucleation hold near 1000°C allows proper development of the internal nuclei, in the case of Minelbite, iron-manganese sulfide. Diopside is the major phase in this glass-ceramic, and it develops on the sulfide nuclei between 850° and 1000°C. Substantial aluminosilicate residual glass remains in the product, which is typically gray.

A white slag-sitall is manufactured in the U.S.S.R. from low-iron (<0.5% Fe_2O_3) slags. This material is formed into sheets with embossed rollers to create patterned wall cladding. It is higher in lime than Minelbite, and is nucleated with a zinc-manganese sulfide. Wollastonite is the major phase, present in ≈1-μm, equiaxial grains. A chemical analysis of this material is presented in Table VIII.

Table VIII. Slag-Sitall Compositions

	U.S.S.R.*	Hungary
	White (wt%)	"Minelbite" Gray (wt%)
SiO_2	55.5	60.9
Al_2O_3	8.3	14.2
CaO	24.8	9.0
MgO	2.2	5.7
Na_2O	5.4	3.2
K_2O	0.6	1.9
ZnO	1.4	
MnO	0.9	2.0
Fe_2O_3	0.3	2.5
S	0.4	0.6
Crystal phases	Wollastonite $CaSiO_3$	Diopside $CaMgSi_2O_6$

*As analyzed at Corning Glass Works.

172

Summary

The major commercial applications for glass-ceramics in the past three decades have been in severe thermal environments (cooking, space, electrical) where the importance of physical and chemical inertness over a wide range of temperatures is critical. The uniquely low thermal expansivity of framework silicates like the lithium aluminosilicates β-spodumene and β-quartz solid solutions and the magnesium aluminosilicate cordierite have allowed the widespread use of glass-ceramics as cookware, high-temperature windows, missile nose cones, regenerators, and heat exchangers. The ability to develop diverse and intricate shapes is also important. The versatility of glass-forming processes can be supplemented by unique secondary finishing in photo-etchable or machinable glass-ceramics, allowing complex dielectric components and unique mechanical features to be produced.

References

[1] S. D. Stookey, *Ind. Eng. Chem.*, **51** [7] 805 (1959).
[2] G. H. Beall, B. R. Karstetter, and H. L. Ritter, *J. Am. Ceram. Soc.*, **50** [4] 181 (1967).
[3] D. R. Stewart, in Advances in Nucleation and Crystallization in Glasses. Edited by L. L. Hench and S. W. Freeman. The American Ceramic Society Spec. Publ., 5, 1971.
[4] G. H. Beall and D. A. Duke, *J. Mater. Sci.*, **4**, 340 (1969.
[5] G. H. Beall and D. A. Duke, Glass, Science and Technology, Vol. 1. Academic Press, New York, 1983.
[6] C. K. Chyung, *J. Am. Ceram. Soc.*, **52** [6] 342 (1969).
[7] W. Ostertag, G. R. Fischer, and J. P. Williams, *J. Am. Ceram. Soc.*, **51** [11] 651 (1968).
[8] D. G. Grossman and J. A. Lanning, *Am. Ceram. Soc. Bull.*, **56** [5] 474 (1977).
[9] W. Schreyer and J. F. Schairer, *J. Petrol.*, **2**, 324 (1961).
[10] D. Lewis III, *Am. Ceram. Soc. Bull.*, **61** [11] 1208 (1982).
[11] S. D. Stookey, *Ind. Eng. Chem.*, **45** [1] 115 (1953).
[12] S. D. Stookey, *Ind. Eng. Chem.*, **46** [1] 174 (1954).
[13] A. I. Berezhnoi; p. 324 in Glass-Ceramics and Photo-Sitalls. Plenum, New York, 1970.
[14] D. A. Duke, J. F. MacDowell, and B. R. Karstetter, *J. Am. Ceram. Soc.*, **50** [2] 67 (1967).
[15] M. J. Buerger, *Am. Mineral.*, **39** [7] 600 (1954).
[16] D. A. Duke, J. E. Megles, Jr., J. F. MacDowell, and H. F. Bapp, "Strengthening Glass-Ceramics by Application of Compressive Glazes," *J. Am. Ceram. Soc.*, **51** [2] 98 (1968).
[17] G. H. Beall et al., *Umschan Verlag*, **72** [14] 468 (1972).
[18] G. H. Beall, in Advances in Nucleation and Crystallization in Glasses. Edited by L. L. Hench and S. W. Freeman. The American Ceramic Society Spec. Publ., 5, 1971.
[19] C. K. Chyung et al, *10th Int. Cong. Glass*, **14**, 33 (1974).
[20] D. G. Grossman, *J. Am. Ceram. Soc.*, **55** [9] 446 (1972).
[21] B. Locsei, pp. 71–74 in Symposium on Nucleation and Crystallization in Glasses and Melts. The American Ceramic Society, Columbus, OH, 1962.

Section **III**

Glass Manufacture

Some Aspects of Tank Melting

J. J. Hammel

PPG Industries, Inc.
Glass Research Center
Pittsburgh, PA   15238

A major step in melting glass is the conversion of batch to molten liquid containing a large fraction of undissolved sand and gaseous inclusions. The most important parameter for rapid conversion of batch to liquid is heat transfer: In a gas-fired furnace, this involves heat transfer to the batch from the flames and radiating walls and crowns, and from the molten glass. This paper discusses the relative importance of batch topography, furnace temperature, and batch chemistry to heat transfer in tank melting and also describes the role of firing and flows in transferring heat from molten glass to the batch blanket.

Melting commercial glasses in large, gas-fired tanks occurs in four steps: (1) conversion of batch to liquid containing a large fraction of undissolved sand and gaseous inclusions, (2) sand dissolution, (3) refining, and (4) homogenization. The most important parameters influencing these steps are time, temperature, and rate of heat transfer. These steps or processes overlap to some extent, and Fig. 1 shows approximate temperatures in large tanks and approximate times required to complete each process. These times are approximated from minimum residence times (MRT) found by previous investigators for container and flat-glass tanks making good-quality glass at 200 to 400 tons per day[1-3]: They found MRTs ranging from 4 to 10 hours. Although completion times for the various processes can vary with throughput, firing rate, profile, etc., the times given are best estimates, based on observations and sampling. For illustration, it is assumed that the total melting time required for all of the steps in Fig. 1 to be completed and provide good-quality glass is approximately seven hours. Although some bubble dissolution and homogenization can occur below 1200°C, it is probably minimal.

The rate of heat transfer is as important as time and temperature in the melting process: Heat transfer from flames, crowns, walls, and electrodes (electric melting and boosting) must be maximized and the batch and glass made good heat receptors. This paper will cover mainly the first step, or approximately the first 45 minutes of the tank melting process, where heat transfer is particularly important.

Conversion of Batch to Liquid

This step includes all of the reactions that occur when bringing batch at room temperature to a molten liquid, containing a large fraction of unreacted sand and undissolved gases, at $\approx 1100°C$. Although the reactions in mixed glass batch are complex, a small pellet (0.3 g) and a Microbar* furnace demonstrate that batch at

---

*ABAR Corp., Feasterville, PA.

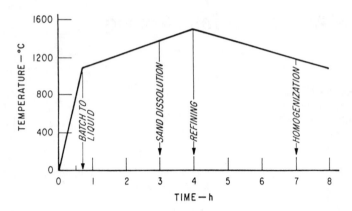

Fig. 1. Approximate times and temperatures to complete melting processes in tank melting.

room temperature can be converted to a molten liquid at 1100°C in less than one minute. Although this demonstrates only the rapidity of batch reactions, other experiments show the highly insulating properties of batch. For example, in one experiment a hemisphere of batch is placed in a furnace at 1540°C and allowed to come to the steady state melting point. Before the batch is placed into the furnace, thermocouples are inserted into the bottom of the batch pile. Figure 2 illustrates measured temperatures at various depths in a melting batch pile. At any given time, the temperatures are approximately 1120°C on the surface, 200°C at 1.3 cm below the surface, and essentially room temperature at 2.6 cm below the surface. Therefore, due to the batch's highly insulating properties, the major problem in glass-batch melting becomes one of heat transfer. Finally, if energy consumption is used as a criterion, calculations show the importance of this first step, in that at 1100°C the melting process has already consumed 85% of the total theoretical heat required, as shown in Fig. 3.

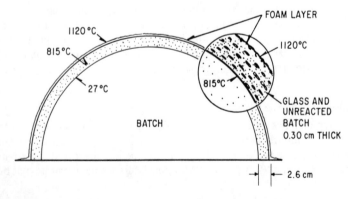

Fig. 2. Cross-sectional view of a melting batch pile at 1540°C furnace temperature.

178

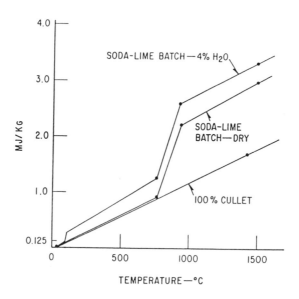

Fig. 3.   Theoretical energies required to melt soda-lime batch and cullet.

Heat Transfer to Batch

Topography

Of all of the factors examined in attempting to improve heat transfer to batch, topography was found to be the most important,[4] as illustrated in Table I. Batch was molded into various shapes (see Ref. 4), and the melting time at 1540°C was measured. Each shape had the same amount of batch and identical base areas. Since the toroid took the least amount of time to melt and did not possess the largest surface area, it was concluded that, although area exposed to flames and radiation is important, the rate of liquid runoff and exposure of cold batch underneath to flames and radiation are more important. This is referred to as "ablation,"[†] since only the surface of the batch is involved in the melting process. This can be

Table I.   Effect of Batch Shape on Ablation Rates

Shape	Surface-to-Volume Ratio	Melting Time (min)	% Rate Increase
Slab (restricted flow)	0.25	49.5	
Slab (unrestricted flow)	0.62	37.3	24
Cone	0.56	36.3	27
Hemisphere	0.50	35.8	28
Logs	0.73	34.5	30
Toroid	0.59	27.3	45

[†]Professor K. Takahashi, now at Okayama University, was the first to recognize the importance of and to develop techniques for measuring ablation rates, at the PPG Glass Research and Development Center.

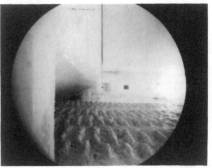

Fig. 4. Batch-blanket configurations with reciprocating tray and Univerbel feeders.

demonstrated readily by placing shaped batch into a furnace and removing it at various times to freeze the liquid layer. Examination of this layer at any given time shows that it is 0.3–0.6 cm thick and contains significant amounts of unreacted sand and bubbles (Fig. 2). Table I also shows that if the liquid formed on the surface of molten batch is not allowed to run off (restricted flow), heat transfer and melting of the underlying batch will be slowed.

Figure 4 shows typical batch-blanket feeding in tank melting. At first, the batch blanket is continuous and composed of many undulations. Farther down the tank, these undulations form logs which produce the liquid runoff required for good melting. Good log formation is mostly a function of the type of feeder (screw, reciprocating, drum-type, etc.) and batch consistency (cohesiveness, etc.). Also, at 100% cullet feeds there is little or no log formation; therefore, good log formation apparently becomes more difficult to obtain as higher cullet percentages are used.

Temperature

Figure 5 shows the effect of furnace temperatures on ablation rates for a typical soda-lime-silica batch: The rates approximately double, going from 12.5 cm/h (68 kg/h/m² or 14 lbs/h/ft²) at 1370°C to 26.3 cm/h (146 kg/h/m² or 30 lbs/h/ft²) at 1590°C. Comparing these changes with the changes in melting times shown on Table I clearly demonstrates the relative importance of temperature and topography.

Heat-transfer rates increase appreciably as the temperature difference between radiating source and receptor increases and distance decreases. Therefore, to increase melting rates, furnace temperatures are maximized (limited by refractories), and temperatures at the surface of the batch blanket are kept as low as possible by good topography and liquid runoff.

Chemistry

Although most sulfate compounds will enhance the melting processes in much the same manner, Na_2SO_4 and $CaSO_4$ are generally added because of compatibility, effectiveness, and price. According to the literature, the chief role of sodium sulfate (salt cake) in melting is for prevention of silica-scum formation,[5,6] acceleration of sand dissolution,[7,8] and removal of gas inclusions.[9,10]

Extensive work at PPG Glass Research and Development Labs showed that salt cake (S/C): (1) enhances ablation rates, (2) accelerates sand dissolution and

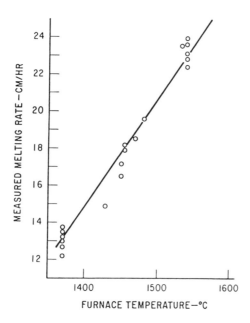

Fig. 5. Batch melting rates as a function of furnace temperatures.

prevents scum formation, (3) reduces melt segregation beneath the batch blanket, and (4) enhances refining rates.

Batch hemispheres with and without S/C were introduced into a furnace at 1540°C and melted. The S/C-containing batch melted 10–15% more rapidly. In addition, partially melted hemispheres were removed from the furnace, and the liquid layers frozen on the batch surfaces were removed and examined: The S/C-containing batch layers were denser and had fewer and larger entrapped bubbles. It was postulated that this allowed more rapid runoff and better heat transfer.

Work at PPG Glass Research and Development Labs has shown that, in both hemisphere and crucible melts, S/C reduces sand agglomeration and segregation early in the melting process. This, in turn, prevents silica-scum formation and accelerates sand dissolution. Karl Bloss[11] showed that during early melting Na_2SO_4 concentrates in rings around a dissolving sand grain. The sulfur profile around a dissolving sand grain was measured with an electron microprobe, and the results are shown in Fig. 6. As shown, the concentration was not maximum at the sand-grain–melt interface, and at least two maxima were found (precipitation in rings, as depicted in Fig. 6). The Na_2SO_4 readily dissolved in the alkali-rich primary melt; the SiO_2 concentration was higher around a dissolving grain, and Na_2SO_4 was less soluble and tended to precipitate. Precipitation occurring in rings is consistent with the Liesegang-rings phenomenon for precipitation first found by Liesegang in 1896.[12] The precipitated sodium sulfate acts as a barrier to particle agglomeration, with some of the precipitated Na_2SO_4 subsequently decomposing and keeping the sand grains dispersed by local agitation, thus preventing scum formation. Previous literature claims a simple wetting mechanism for dispersion,[13–15] which suggests a high concentration of Na_2SO_4 at the sand-grain–melt interfaces. This is not consistent with our experimental findings.

SULFUR PROFILE

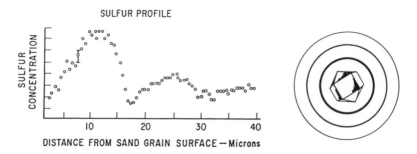

DISTANCE FROM SAND GRAIN SURFACE — Microns

Fig. 6. Electron microprobe analysis and conceptual view of sodium sulfate precipitation around a dissolving sand grain.

Merritt Hummel[16] developed a test that gives information on melt segregation beneath a melting batch blanket: Batch is placed on molten glass and melted for a specified time, then removed from the furnace and cooled. Figure 7 shows the results of two independent tests, at 1290°C, at S/C levels of 0.45, 2.25, 4.5, and 9 kg per 455 kg of sand for a soda-lime-silica batch. Stringers and droplets of inhomogeneous glass, 2–10 mm in diameter, form beneath the batch blanket.

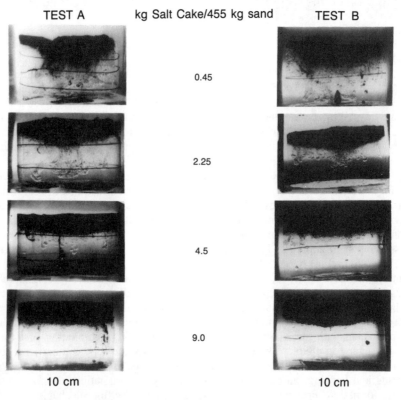

TEST A   kg Salt Cake/455 kg sand   TEST B

0.45

2.25

4.5

9.0

10 cm   10 cm

Fig. 7. Change in melt segregation beneath a melting batch pile as a function of salt-cake levels in the batch.

Table II. Composition and Properties of Molten Droplets beneath a Melting Batch Pile

	Electron Microprobe Analysis, 4.5 kg (10 lbs) S/C at 1290°C	
	Base Glass, %	Droplet, %
SiO_2	73.2	62.9
Na_2O	13.8	14.5
CaO	8.9	14.3
MgO	4.0	6.0
Calculated		
Density	2.49 g/cm³	2.66 g/cm³
Log_2 viscosity	1443°C	1273°C

Electron microprobe analyses of the stringers and droplets show them to be alkali- and alkaline earth-rich (Table II). Table II also shows that the droplets are denser than the base glass, therefore causing a gravity separation. Increasing S/C levels reduces melt segregation, with the least segregation found at 9 kg S/C per 455 kg of sand.

Heat Transfer from Molten Glass to Batch

A certain fraction of the batch-blanket melting occurs from below. This is generally brought about by transferring heat to the melt, down tank of the batch and foam lines and pumping it back beneath the batch blanket. Melting of batch in contact with the molten glass is estimated to equal one-quarter to one-third of the total blanket melting, depending on the firing profile and resulting tank flows. This fraction is not surprising since, in practice, the temperatures of the energy sources at the top and bottom of the batch blanket are quite different. Generally, the melting rate from below is slower than that from above. However, there are definite advantages in transferring heat directly to the melt down tank — one is development of the required flows for good-quality glass.

Firing and Flows

Properties of the batch that make it a good heat receptor for rapid melting have been covered; properties of flames that maximize heat transfer to the batch and the role of the flows in melting will now be discussed.

Efficiencies would be highest if it were possible to transfer heat directly from the flames to the batch. Since this transfer is not practical, much of the melting energy is obtained from crown reradiation. However, by obtaining best flame properties, it is possible to maximize heat transfer directly from the flame to the batch blanket. Flame temperature should be as high as practical, the limit being the temperature stability of crown and regenerator refractories. This will increase ΔT between the flame and batch and improve radiant-heat transfer. Standard techniques used for many years are regeneration and recuperation. These techniques not only recover energy by preheating combustion air, but also significantly increase flame temperature: from 1900°C flame temperature without preheated air to 2600°C with combustion-air preheat to 1200°C.

The flames should be broad, of relatively low velocity, and as close to the batch as possible for maximizing the flame-covered area and improving both radiant and convective heat transfer (Fig. 8). Allowing the flames to run relatively parallel to the batch blanket avoids impingement, thus reducing batch carryover

Fig. 8.   Typical flame configuration in tank melting.

and related refractory wear and regenerator plugging, while still allowing good heat transfer.

For highest efficiencies, the fuel (natural gas, producer gas, or oil) should be well mixed to allow low excess air while obtaining good combustion (low combustibles). Most tanks are now equipped to fire with both gas and oil because of availability and cost problems. Although flame properties differ between gas and oil, no major differences are apparent in batch melting, so that it would be difficult to choose one over the other in terms of melting improvements.

As mentioned previously, tank flows play a major role in determining the rate and completeness of the melting processes, since they determine total melting time and temperature for any volume element of batch or molten glass. In most tanks, cold batch is fed at one end (see Fig. 9), with varying rates of overhead firing from the individual ports as the batch blanket moves downtank. Thus, relatively cold molten glass produced at the feed end gradually increases in temperature as it approaches the "hot spot" downtank. The exact location of this hot spot depends mainly on feed rate and firing profile. Since the hotter glass is less dense, the molten glass has a slightly higher depth in this zone, producing longitudinal flows back to the feed end, as depicted in Fig. 9. This upwelling is generally referred to as the "spring zone." In like manner, lateral flows are set up due to the cooling effect of the furnace walls.

In tank melting, a strong spring zone (large ΔT, with accompanying good rearward flow) is important for efficiently accomplishing all four steps in the melting process. This rearward flow of surface glass produces resistance to forward

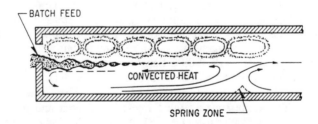

Fig. 9.   Cross section of longitudinal flows in conventional tank melting.

flow of the batch blanket and thus establishes a natural barrier to free movement of unmelted batch downtank. A strong return flow can actually cause unmelted batch to move back to the feed end. This resistance to forward flow increases the residence time of the batch blanket in the "melter" portion of the tank. A strong return flow of hot molten glass also produces good heat transfer and melting of the batch from below.

Acknowledgments

I appreciate the opportunity given by PPG Glass Research and Development Labs to work on the scientific, as well as the technological, aspects of glass melting and wish to acknowledge J. E. Cooper, L. J. Shelestak, M. J. Hummel, and K. H. Bloss of PPG Glass Research and Development for their valuable inputs.

References

[1]R. F. Barker, "A Radioactive Method Suitable for the Study of Mass Flow Rates Through a Large-Scale Glass-Melting Tank," *J. Soc. Glass Technol.*, **42**, 101–108 (1958).

[2]J. J. Smith and T. E. McGary, "Tracer Studies with Samarium via Neutron Activation Analyses," *Glass Ind.*, **49**, 78–84 (1968).

[3]V. G. Leyens, J. Smreck, and J. Thyn, "Use of Isotope Measurements and a Mathematical Model to Determine the Residence Time Distribution in a Glass Tank," *Glastech. Ber.*, **53**, 124–29 (1980).

[4]J. J. Hammel and J. D. Mackenzie, "Glass Melting Enhancement by Toroidal Batch Shaping," U.S. Pat. No. 4 282 023, August 4, 1981.

[5]A. Dietzel and O. W. Flörke, "Action of Sulfate in the Melting Process," *Glastech. Ber.*, **32**, 181–85 (1959).

[6]T. Inoue, "Behavior of Residual Sand Grains in the Glass Melting Process," Symposium on Defects in Glass, Tokyo, Japan, 1966.

[7]J. E. Stanworth and W. S. Turner, "The Effect of Small Additions of Sodium Sulfate on the Reactions in the Mixture $6SiO_2 + Na_2CO_3 + CaCO_3$," *J. Soc. Glass Technol.*, **21**, 359–67 (1937).

[8]L. Nemec, "A Contribution to the Study of Glass Melting and Refining"; pp. 155–65 in XIth International Congress on Glass, Vol. 4, Prague, Czechoslovakia, 1977.

[9]C. Thorpe, "The Effect of Small Additions of Sodium Sulfate on the Refining of Glass," *Glass Technol.*, **3**, 135–39 (1962).

[10]F. Shaw and S. P. Jones, "Effect of Sodium Sulfate and Furnace Atmosphere on Refining Container Type Glass," *Am. Ceram. Soc. Bull.*, **45**, 1004–1008 (1966).

[11]K. Bloss, PPG Glass Research and Development Labs; private communication.

[12]S. Glasstone; pp. 1264–66 in Textbook of Physical Chemistry. Van Nostrand, New York, 1954.

[13]A. R. Conroy, W. H. Manring, and W. C. Bauer, "The Role of Sulfate in the Melting and Fining of Glass Batch," *Glass Ind.*, **47**, 84–89, 133–39 (1966).

[14]I. Tokuhara, S. Iseki, and T. Fujiwara, "Studies of the Effect of the Addition of Sulfates and Sulfites to Glass Batches"; pp. 20–23 in Xth International Congress on Glass, Vol. 3, Kyoto, Japan, 1974.

[15]Y. Kokubu, J. Chiba, and T. Okamura, "The Behavior of Sodium Sulfate during the Glass Melting Process"; pp. 147–54 in XIth International Congress on Glass, Vol. 4, Prague, Czechoslovakia, 1977.

[16]M. Hummel, PPG Glass Research and Development Labs; private communication.

Quality Control in the Glass-Container Industry

JOHN S. WASYLYK

American Glass Research, Inc.
615 Whitestown Road
Butler, PA   16003-0149

Quality- and process-control methods and procedures for determining finished ware quality in today's glass-container industry are reviewed in this paper. Manual ware-sampling, inspection, and test procedures are compared with newer, similar automated procedures based on an individual-cavity-identification concept.

Improvements in glass-container melting and production technology have resulted in increased container-production speeds that would have been thought impossible only a few years ago. Such increased production rates have also increased demands on inspection, testing, and quality control in modern container-production plants. Our intent is to outline present and future quality- and process-control methods.

Container-Production Process

Modern glass containers are typically made on high-speed, continuous-production, individual section (IS) machines. Glass containers are produced on a 24-hour-a-day, 7-day-a-week, 52-week-a-year schedule. Each IS machine is comprised of from as few as 4 to as many as 10 individual sections; each individual section is composed of from 1 to 4 mold sets or cavities. A cavity consists of two paired molds, called, respectively, the blank, or parison mold and the blow mold.

Bottle-production rate depends primarily on the glass weight and the size of the particular bottle being formed. Production rates vary from as few as 3 to 4 cavities per minute for larger bottles, to as many as 12 to 14 or more bottles per minute for smaller or lighter production. Overall mechanized production rates, therefore, might vary from 12 to over 400 bottles per minute, or from 120 to more than 4000 gross of bottles per day. The number of bottles produced in a day obviously increases as machine size increases and/or bottle weight decreases.

Present-day manual inspection and testing functions may also become taxed as production rates increase. Bottles must now be sampled on an individual-cavity basis, and the sampled bottles must be tested in various ways; for example, dimensional and performance tests must be performed. Those test results must be interpreted and correlated with each other and with the other cavities on each production machine. The data must then be compared with previous production data to accurately and confidently determine whether significant quality shifts have occurred. The data must also be reported to the hot-end machine operator for correction of undesired changes in the production operation.

All of the previously mentioned sampling and inspection procedures traditionally have required significant time. They have also required intensive manual labor, which is a significant cost factor in modern container production. Despite intensive efforts at reducing the many unit-container production costs in recent years, labor remains a significant production-unit cost factor.

The time lag involved in manual sample acquisition, testing, and reporting of the properly formatted data to the production unit for corrective action is also critical, and even more so if a genuine process-control function is to be provided to the container-production system.

Since containers are produced in individual sets of molds or cavities, and since, to a large degree, the quality of containers produced in one set of molds is independent of the quality produced in an adjacent set of molds, production from each cavity must be sampled and inspected on a specified time schedule.

Past and Present Quality- and Process-Control Procedures

Glass containers must meet a variety of specifications to be considered of commercial quality. Those specifications apply to many bottle dimensions and capacities, and to various aspects of glass- and bottle-forming quality.

The most stringent performance requirements for glass containers are generally those established for pressure ware, i.e., containers intended to convey carbonated products to the consumer. Since this class of container must frequently withstand simultaneously applied loads of internal pressure, thermal shock, vertical load, and impact, various dimensional and performance specifications have been developed for them. Conformance to those specifications is established by test procedures that typically are carried out during the manufacturing operation.

The Voluntary Product Standard, or VPS,[1] is administered by the National Bureau of Standards. The VPS is a set of manufacturing requirements used as voluntary production standards by all U.S. glass-container manufacturers.

The National Soft Drink Association also has published a series of voluntary specifications[2] that apply to the manufacture of carbonated-soft-drink containers.

Containers intended for carbonated beer are sampled and tested under the aegis of the United States Brewers Association guidelines,[3] promulgated in 1970. The USBA specification guidelines follow the same ASTM test methods as do the VPS.

The VPS cites a series of standard methods from the American Society for Testing and Materials[4] for sampling and testing containers intended for carbonated beverages. The ASTM procedures are, respectively: ASTM C–224, "Standard Method of Sampling Glass Containers"; ASTM C–147, "Standard Methods of Internal Pressure Test on Glass Containers"; ASTM C–148, "Standard Methods of Polariscopic Examination of Glass Containers"; and ASTM C–149, "Standard Method of Thermal Shock Test on Glass Containers."

According to the VPS, NSDA, and USBA requirements, containers must be sampled from production according to ASTM C–224, "Sampling Procedures for Glass Containers." The sampling of containers in the ASTM C–224 method is based, in part, on the sampling procedures detailed in the MIL–STD–105D procedure.[5] That set of procedures is predicated on sampling by attributes from a process producing discrete series, or lots, of product. Various sampling plans have been established, based on the size of the production lot being evaluated.

The above sampling and testing methods generate data in both attribute and variable formats. Attribute data consist of go/no-go information, in which bottles must meet an established criterion. Variable performance data, i.e., data which express quantitative results, arise from performance testing, such as internal-pressure or thermal-shock testing, or from dimensional information, such as bottle-sidewall thickness, height, or diameter. Both sets of data are compared with specification values for each of the attributes or variables being evaluated; a decision is then made to accept the production up to that time, based on the data, or to reject the production until specifications are met.

Following data acquisition and the acceptance or rejection of production, decisions must be made affecting the continued course of production: To whom must the data be reported? Is a change in the process warranted? Which parameters must be changed? In which direction must the changes be made, and of what magnitude?

The data are also reported to the hot-end production units. The times between which the data are generated and reported become of prime importance for process corrections. Unwarranted process corrections must not be made, nor should necessary corrections be ignored.

The heart of any effective process-control system, applied to glass-container production, then requires some means (1) of individually identifying the cavity or particular set of molds from which each container is produced and (2) for correlating the various inspection and performance test data from each of the tests with the cavities from which the bottles were produced.

Increased Present and Future Quality- and Process-Control Needs

In the past, many inspection functions were performed manually. Bottles were removed by hand from the production stream and carried to the quality-control-testing location or laboratory, where various dimensional and performance characteristics were determined. Following assembly of the appropriate information, the data were then reported to the personnel in the manufacturing plant responsible for corrective action, if such measures were warranted. Such manual procedures were tedious and subject to error. There was also a considerable time lag between the onset of a production problem and the detection of that problem.

A plug gauger was one of the earliest automated inspection devices installed on a container-production line. A plug gauger consists of a short, nonmetallic plunger accurately sized to the minimum desired inside diameter of the bottle finish. The plunger is automatically inserted into the finish of each bottle in the production stream to ensure the adequacy of the inside-finish dimension, and bottles with insufficient inside-finish diameters are rejected. The cavities producing undersized finishes must be manually determined before any action can be taken to eliminate the dimensional problem.

Another automatic online inspection device is the finish-check detector. Bottles are rotated in the check detector while a narrow, collimated beam of light is directed against the bottle finish at a grazing angle. A light-sensitive photocell is angled to the incident light beam, so that if the finish contains no checks or cracks to reflect the light beam the photocell sees no light and the bottle is passed. If a check large enough to cause reflection is present in the finish, the photocell triggers a rejection mechanism, removing the affected bottle from the production stream. Light-source/photocell combinations are positioned and stacked in the check detector to cover various areas on the container surface.

However useful the check detector may have been in eliminating bottles with checks, its major drawback was in not providing immediate information to the production department. The cavity or cavities corresponding to the ware being rejected still had to be manually determined. Thus, check detectors provided a means of rejecting substandard ware, but were inefficient for providing immediate and sufficient information on eliminating the problem.

Fig. 1. Bottle exiting AGR automatic online thickness-inspection device, showing three separate wall-thickness-sensing heads.

The next significant improvement in container production came in the form of automatic wall-thickness-gauging equipment. Maintenance of adequate sidewall thickness is important in many container applications, especially in the pressure ware. Previously sampled, manual laboratory measurements were replaced in the early 1960s with automatic online wall-thickness determinations using a basic capacitance principle: An upright bottle, rolled along a linear capacitance-sensing head, comprises the dielectric element in the assembly, as seen in Fig. 1. The wall thickness of the bottle at each contact point on the sensing head is electronically determined as a capacitance value, which, in turn, converts to a glass-wall-thickness value. If the minimum wall thickness sensed during a complete rotation of a given bottle circumference over the sensing head falls below the minimum desired value, that bottle is automatically rejected. As many as four sensing heads may be positioned along the cylindrical sidewall height of a bottle to allow determinations of wall thickness at each location.

Once again, as with the advent of the check detector, significant improvement was obtained in detecting and eliminating problem ware. But the affected cavity or cavities related to rejected ware still had to be manually identified and appropriate corrective action taken.

In the early 1970s, American Glass Research, Inc. (AGR) began work on a cavity-identification system allowing automatic identification of the cavity in which a given bottle was produced. The present cavity-identification system involves identification of a series of hemispheres, molded into the outside bottom surface of the bottle. The hemispheres, or dots as they are commonly called, are

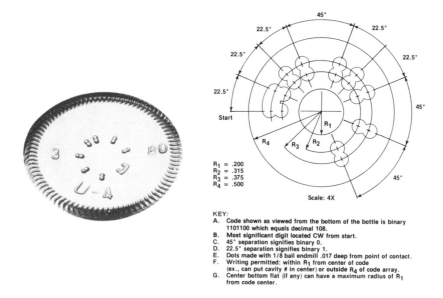

R₁ = .200
R₂ = .315
R₃ = .375
R₄ = .500

Scale: 4X

KEY:
A. Code shown as viewed from the bottom of the bottle is binary
   1101100 which equals decimal 108.
B. Most significant digit located CW from start.
C. 45° separation signifies binary 0.
D. 22.5° separation signifies binary 1.
E. Dots made with 1/8 ball endmill .017 deep from point of contact.
F. Writing permitted: within R₁ from center of code
   (ex., can put cavity # in center) or outside R₄ of code array.
G. Center bottom flat (if any) can have a maximum radius of R₁
   from code center.

Fig. 2. AGR cavity-identification dot code, showing details of individual dots and arrangement on bottle outside bottom surface.

arranged in a circumferential pattern, as seen in Fig. 2.

The presence of the dots and their relative positions may be used to encode certain bottle-identification information. The seven-bit code used can encode a maximum of 128 separate numbers or characters in that portion of the bottle bottom assigned to the dot coding. The dot patterns are read by a sophisticated solid state array camera. The binary-coded information in each bottle is interpreted by a dedicated system computer; the dot code on every bottle bottom in the production stream is thus read and identified by the cavity-identification system.

Originally, the cavity-identification system was intended to serve as a cavity-identification and bottle-selection and rejection device. Bottles from desired cavities could be selected on a sampling basis for later testing, or all production from a given cavity or cavities could be rejected, all automatically by the cavity-identification unit.

However, the evolutionary process took another significant step when the cavity-identification unit was interfaced with the online wall-thickness selector, a process-control or host computer, and an automatic tracking system. This system allowed an individual bottle to be tracked through the entire online inspection system. The interfacing allowed the correlation of online wall-thickness data with a particular cavity, and cavity-specific wall-thickness data could be instantly and directly displayed in the production-plant hot end. Such displays provide timely and efficient information to production personnel, allowing proper corrective action to be taken. The addition of the host computer, as well as the development of appropriate equipment interfaces, allowed check detectors and plug gaugers to be interfaced with the cavity-identification system, as seen in Fig. 3. Those interfaces provided even more timely and useful information to production personnel.

As long ago as the late 1930s, Preston[6] demonstrated the variability inherent in internal pressure-strength data, as sampled directly from a production line (see

191

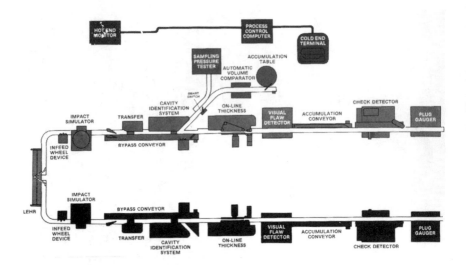

Fig. 3. Schematic of AGR cavity-identification, sampling, and online-inspection equipment, with host computer, display monitor, and control terminal.

Fig. 4). Teague and Blau[7] also showed similar variability in their classic work on the internal pressure strengths of glass bottles, as shown in Fig. 5. Production-plant operating personnel must now extract the information necessary for efficient production control from similar, sampled internal-pressure-strength variable data.

Various procedures for establishing control charts for mean density, density range, and density standard deviation to control glass composition have been published by Ghering.[8] However slow such density changes may be, the directions of density shift may be predicted fairly well from prior density data. Changes in internal pressure strengths may be more rapid and the causes for such changes more varied, making internal pressure strengths less predictable, due to both process variability and statistical-analysis limitations.

Since the development of both the AGR automatic-sampling internal-pressure tester and the AGR automatic-sampling volume comparator, sampled variable data can be correlated with the cavity in which the respective bottles were manufactured, as are cavity-correlated minimum-wall-thickness data. Other manufacturers' automatic online bottle-dimensional-gauging equipment can also be interfaced with the developing AGR automated process-control system, as can other cavity-correlated data relating to bottle-checking behavior.

Improved and more efficient statistical techniques must be developed for handling and analyzing the available cavity-correlated data accurately, and confidently signaling significant process changes to the appropriate production units for correction.

Summary

Improved sampling and performance-test techniques must allow functional processing of the tremendous quantities of data that are possible from modern inspection devices and process-control systems now being developed. The reduction of such masses of data into useful formats for personnel controlling the modern

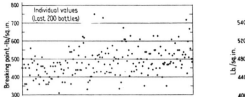
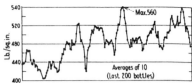

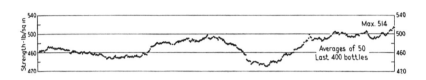

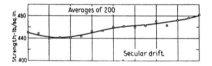

Fig. 4. Various units for moving average internal-pressure-strength data for bottles sequentially sampled directly from production line (Ref. 6). Larger-unit moving-average data show a shift in internal pressure strength for the particular process sampled.

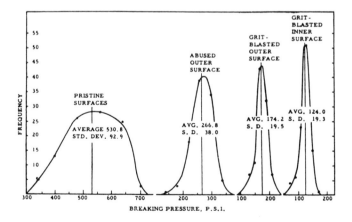

Fig. 5. Variability of internal-pressure-strength distributions for pristine bottles sampled directly from production by Teague and Blau (Ref. 7), compared with variability in pressure strength for purposely abused ware.

193

container-production process is also of prime importance. Future lightweight containers will further increase the need for more efficient sampling, inspection, and data-reporting methods. Using data generated by an automated sampling and inspection system, in a closed-loop feedback control system, to automatically regulate container production is an extension of present quality- and process-control developmental efforts, and will become a working reality in the very near future.

References

[1] Voluntary Product Standard, ANSI/VPS PS 73–77, Carbonated Soft Drink Bottles; available from the Standards Development Services Section, National Bureau of Standards, Washington, DC 20234.

[2] Bottle Specifications Guideline, November 1970; issued by the National Soft Drink Association, 1101 16th Street, NW, Washington, DC 20036.

[3] Brewing Industry Recommended Beer and Ale Bottle Purchase Specifications Manual; issued by the United States Brewers Association, Inc., 1750 K Street, NW, Washington, DC 20006.

[4] American Society for Testing and Materials, 1916 Race Street, Philadelphia, PA 19103.

[5] Military Standard, Sampling Procedures and Tables for Inspection by Attributes, MIL–STD–105D, 29 April 1963; available from the Naval Publications and Forms Center, 5801 Tabor Avenue, Philadelphia, PA 19120.

[6] F. W. Preston, "Concerning the Strength of Weakest Bottles," *J. Am. Ceram. Soc.*, **20** [10] 329–36 (1937).

[7] J. M. Teague and H. H. Blau, "Investigations of Stresses in Glass Bottles under Internal Hydrostatic Pressure," *J. Am. Ceram. Soc.*, **39** [7] 229–52 (1956).

[8] L. G. Ghering, "Refined Method of Control of Cordiness and Workability of Glass during Production," *J. Am. Ceram. Soc.*, **27** [12] 373–87 (1944).

Proof Testing — A Tool for Quality Assurance

KARL JAKUS

University of Massachusetts
Amherst, MA   01003

This paper reviews the principles of proof testing glasses and ceramics and discusses the benefits and limitations of practical proof-test methodologies. Proof testing is shown to be beneficial for improving the reliability of glass and ceramic components, provided that fatigue during the proof test is eliminated or at least minimized. Existing theory can predict the results of proof testing in inert environments and environments where fatigue plays a relatively minor role. This theory, however, is inadequate for predicting results of proof testing in environments with severe fatigue effects.

Proof testing is intended to ensure, or at least improve, the probability of survival for objects in specific applications. In a broad sense of the word, proof testing is commonly practiced every day, either intentionally or unintentionally. For example, I usually perform a simple "test" before stepping on unproved ice: I stomp or push on the ice with one foot before putting my weight on it. If the ice breaks through, I retreat. If the ice passes my "test" I take my chances and proceed. This is intentional proof testing. On the other hand, an example of unintentional proof testing is when eggs are collected from the hens on a poultry farm. During collection the eggs are bumped and jounced a lot more than they will ever be once they are put in cartons. The weak eggs break and obviously are rejected; the survivors will surely make it to market. The underlying principle in this sort of proof test is intuition and experience, both of which are important, but not sufficient, for industrial application. In that case, one must quantitatively know the reliability of the proof-test survivors or, in other words, be able to calculate the benefit of proof testing for a particular product.

Glass and ceramic products, brittle as they are, can particularly benefit from proof testing when they are to be used for load-bearing. In fact, proof testing has proved to be one of the most reliable quality-assurance tools for products made of these materials. This technique's usefulness was recognized long ago and, accordingly, clever and very beneficial empirical proof-test procedures were developed by numerous industries to assure the reliability of their own products. A case in point is the "overspin" test for wheels in the grinding-wheel industry, developed to minimize the safety hazard associated with spontaneous wheel bursts. These empirical techniques are usually product specific and not readily transportable from product to product, even if differences are relatively minor. Fortunately, a quantitative understanding of the failure mechanisms in glasses and ceramics evolved in the 1960s, laying the foundation for developing a rational proof-test methodology. Accordingly, in the early 1970s a proof-test theory was formulated by such researchers as A. G. Evans, S. M. Wiederhorn, and E. R. Fuller, Jr., among others.[1-3] This paper reviews the principles and salient features of this theory and discusses the benefits, as well as the limitations, of practical proof testing.

Basic Principles

The fundamental idea of proof testing for improving strength is intuitively simple: Each component is put under excess stress to break the weak specimens and retain those that are stronger than the applied proof stress. This, of course, assumes no damage to the components during the proof-test process — aside from breaking the weak ones — so that the survivors remain as strong as they were before the test. Such a selection process statistically improves the strength of the component population as a whole. Another crucial but somewhat hidden assumption in this rudimentary model of proof testing is that the proof stress is exactly what the components will endure in service. It is not difficult to see that if the latter assumption is violated the proof test will not show any benefits. On the other hand, it is not as easy to see what happens when some strength degradation occurs during the proof test: In this case, whether the test will or will not result in a statistically improved population depends on the degree of damage. This aspect of proof testing must be examined most carefully in the case of glasses and ceramics, since they typically exhibit some strength degradation when subjected to stress. Whether the applied stress is a service or a proof stress is irrelevant.

According to current understanding, glasses and ceramics fail due to pre-existing flaws on or within the specimen under stress. In certain environments these preexisting flaws may grow in a stress-dependent fashion until they reach a critical size, at which time spontaneous fracture occurs. Thus, a component may withstand a certain load for a while, but could fail later if the stress is sustained. The mathematical model for this failure process was developed in terms of fracture mechanics.[4,5] It is assumed that flaws in brittle materials are sharp cracks that grow as a function of the instantaneous stress-concentration factor K_I. The appropriate stress-concentration factor for such a flaw is

$$K_I = \sigma Y \sqrt{a} \tag{1}$$

where σ is the applied stress, a the flaw size, and Y a shape factor which equals $\sqrt{\pi}$ for a half-penny crack. When K_I reaches a critical magnitude, the specimen fails; the applied stress at that instant is called the strength, S.

The simplest and most widely accepted model for the growth of this crack is a power law in K_I,

$$v = da/dt = A(K_I/K_{IC})^N \tag{2}$$

where K_{IC} is the critical stress intensity factor, t is time, and A and N are environmentally dependent constants.

Substituting Eq. (1) into Eq. (2) gives a differential equation that can be integrated, in principle, for any applied-stress history. The result shows the strength degradation that occurs from the onset to the termination of stress. For an arbitrary variation of applied stress with respect to time, $\sigma(t)$, this integration takes the form

$$S^* = \left[(S_i^*)^{(N-2)} - C \int_0^{t_f^*} \sigma^{*N} dt^* \right]^{1/(N-2)} \tag{3}$$

where S^* is the strength after the proof test, S_i^* the initial strength, and t_f^* the time at fracture for those specimens failing during the proof test and the time at the end of the proof test for the surviving specimens. Furthermore, C is a constant defined as

$$C = [t_0 S_0^2 (N - 2) A Y^2]/2 K_{IC}^2 \tag{4}$$

It must be pointed out that the strengths and stresses (S^*, S_i^*, and σ^* in Eq. (3)) are normalized with the characteristic strength of the initial population, S_0 (typically the Weibull shape parameter), and the times are normalized with the total proof-test duration t_0.

For a triangular proof-stress pattern, that is, constant loading and unloading stressing rates, Eq. (3) can be integrated in closed form.[6] The most interesting result of this integration is the formula for the residual strength, S_r^* of the specimens surviving the test, namely

$$S_r^* = \left[S_i^{*(N-2)} - C\frac{\sigma_p^{*N}}{N+1} \right]^{1/(N-2)} \tag{5}$$

One can also calculate the residual strength, S_r^*, of the specimens that just pass the proof test,[6] i.e.,

$$S_{r_{min}}^* = [(N-2)\,\sigma_p^{*N}/C(1-t_p^*)]^{1/3}(3/N+1)^{1/(N-2)} \tag{6}$$

where t_p^* is the normalized time at the peak proof stress σ_p^*.

In many practical applications one assumes that the initial strength distribution can be represented by a Weibull function,

$$Q_i = (S_i/S_0)^m = (S_i^*)^m \tag{7}$$

where $Q_i = \ln(1/1 - F_i)$, F_i is the cumulative failure probability, and m and S_0 are the Weibull shape and location parameters, respectively.

Substituting Eqs. (6) and (7) into Eq. (5) gives an equation for the residual strength in terms of failure probabilities,

$$S_r^* = \left[(Q_f + Q_p)\frac{N-2}{m} - (Q_p)\frac{N-2}{m} + (S_{r_{min}}^*)^{(N-2)} \right]^{1/(N-2)} \tag{8}$$

Equations (1) to (8) represent the backbone of the mathematical framework for proof testing glass and ceramic components. At this point it is prudent to reiterate the assumptions in this theory, for a clear understanding of its applicability.

1. Specimens fail due to preexisting flaws.
2. The flaws are cracklike in nature.
3. The flaws grow according to a power law in K_1 whose constants depend on the environment.
4. The initial strength distribution is represented by a Weibull function.
5. The equations in this paper were derived for a triangular proof-stress pattern.

It should also be reemphasized that for the proof test to improve product reliability the proof stress must represent the actual service stress.

Casual observation of the equations does not indicate which parameters govern such important practical considerations as how many specimens break during the proof test, the minimum strength of the survival population, and which proof-test methodology (environment, loading, and unloading rates) gives best results in a given situation. For greater insight into the effects of the variables, the equations were evaluated for several typical situations, along the lines of a recent paper by Fuller et al.[6] The results of these calculations are discussed next.

Evaluation of the Equations

A useful way of illustrating what happens to a specimen's strength during proof testing in a fatigue environment is by plotting Eq. (3) as a function of time.

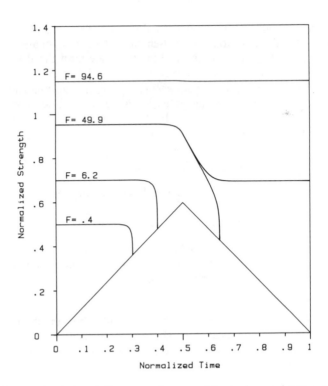

Fig. 1. Strength degradation map for proof-tested specimens: $m = 7.7$, $S_0 = 126.9$ MPa, $N = 17.34$, $B = 0.2794$ MPa$^2 \cdot$s, $\sigma_p^* = 0.60$, $\dot{\sigma}_u^* = 1$ s^{-1}, $\dot{\sigma}_d^* = 1$ s^{-1}.

This plot was made for a hypothetical proof-test cycle, and the results are shown in Fig. 1. Here, the Weibull constants ($m = 7.7, S_0 = 126.9$ MPa) and the fatigue parameters ($N = 17.34, B = 0.2794$ MPa$^2 \cdot$s) were chosen to correspond to typical test conditions for soda-lime glass in a moist air environment. The triangular proof-test cycle with a normalized peak stress of 0.6 was chosen to break 50% of the specimens during the proof test. The figure shows the proof-test triangle and strength histories of five specimens. It can be seen that those specimens with an initial strength below a critical value ($S_i^* = 0.95$, in this case) failed during either the loading or the unloading segment of the proof-test cycle. Specimens with an initial strength above this threshold survived the test, although they may have suffered substantial strength degradation in the process.

Specimens with initial strengths slightly above the fail–survive boundary are the most critical with respect to the minimum-strength level of the after-proof population. These specimens can suffer such a large degree of strength degradation that their residual strength may fall substantially below the peak proof stress, and in many cases this strength becomes essentially zero. Equation (5) gives the value for the residual strength of a specimen that just passed the proof test. In this example, the normalized minimum strength was 0.003, which can be considered zero for practical purposes. Although ending with no or very low assured minimum strength detracts from the value of proof testing, there is some solace for the application engineer despite this undesirable fatigue effect: The number of speci-

198

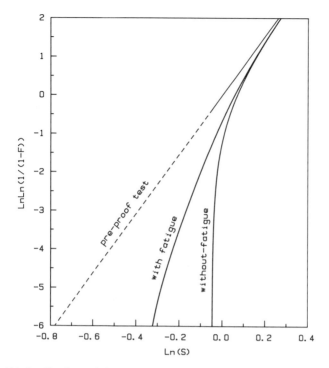

Fig. 2. Weibull plot of initial and after-proof strength distributions:
$m = 7.7$, $S_0 = 126.9$ MPa, $N = 17.34$, $B = 0.2794$ MPa$^2 \cdot$s, $\sigma_p^* = 0.60$,
$\dot{\sigma}_u^* = 1$ s^{-1}, $\dot{\sigma}_d^* = 1$ s^{-1}.

mens ending with residual strengths below the peak proof stress is typically low. In the example at hand, only 14 of 1000 specimens had such low strength. Therefore, a minimum strength cannot be ensured, but the probability that any specimen will be weaker than the peak proof stress is very low.

Another customary way to show the effects of proof testing is by plotting Eq. (5), the residual strength, on Weibull coordinates, as shown in Fig. 2. This figure assumed the same test conditions as in Fig. 1. The three curves in the figure represent initial strength distribution (a straight line on the Weibull plot), after-proof strength distribution with fatigue, and after-proof strength distribution without fatigue. The latter curve represents the results of a proof test performed in an inert environment. In this example, proof testing resulted in improved surviving populations, although the distribution without fatigue was markedly superior to that with fatigue. Specifically, the distribution without fatigue asymptoted to the peak proof stress, in effect the guaranteed minimum strength, whereas the distribution with fatigue asymptoted essentially to zero at very low failure probabilities, thus offering no assured minimum strength. It should be pointed out that in Fig. 2 the peak proof stress with fatigue was chosen to be lower than that without fatigue (σ_p^* (fatigue) = 0.6, σ_p^* (no fatigue) = 0.95) so that the component loss due to proof testing would be the same (50%) in both cases. The dashed segment of the pre-proof test curve on the figure represents specimens that failed during either of the proof tests. If the peak proof stress without fatigue were set to the lower value of 0.60, only about 1.8% of the specimens would break. On the other hand, if the peak proof test with fatigue were set to the higher value of 0.95, used for the inert case, virtually all of the specimens would fail. In summary, fatigue increases specimen loss during proof testing and decreases the assured minimum strength of the surviving population; in fact, this minimum strength often approaches zero.

What can be done to reduce component loss and improve the statistical strength of the after-proof population? Fatigue can be eliminated or minimized during proof testing, either by removing moisture from the test environment using vacuum, oil, or another moisture-free medium, or by performing the test at a very low temperature (in liquid nitrogen, for example), where the rate of the crack-growth reaction is substantially reduced. If fatigue-free conditions cannot be achieved, as in many industrial situations, the proof test should be performed in as dry an environment as possible. Testing in dry air, compared to moist air or water, reduces component loss and increases the after-proof strength at all levels of failure probability, although the assured minimum strength may still be quite low.

Examining Fig. 1 leads to the conclusion that very rapid unloading would allow no time for strength degradation after the proof stress reached its peak value. Minimizing fatigue this way decreases the number of specimens lost in the proof test but does not substantially increase the predicted minimum strength or the statistical strength of the surviving population. This technique simply rescues those specimens whose initial strength falls in the close proximity of the fail–survive boundary. Of course, if the unloading were truly instantaneous, the minimum strength would equal the peak proof stress, since the strength of the specimens passing over the peak proof stress could not be further degraded. In our example, if the unloading rate were increased 100-fold the number of specimens failing during the proof test would decrease from 50 to 40%, but the assured minimum strength, S_r^* would only increase from 0.003 to 0.015, still too small for practical significance. It should be noted that the loading rate was the same in both of these proof-test calculations; only the unloading rate varied.

The number of specimens broken during the proof test can be substantially reduced if the entire proof-test cycle (loading plus unloading) is shortened. In our example, if this total duration were reduced 100-fold, only 6.6% of the specimens would break in the test. Unfortunately, the predicted minimum strength would still rise very little. Thus there seem to be practical ways of reducing component loss during the proof test, but only eliminating the fatigue effect will guarantee a minimum after-proof strength. If fatigue during the proof test is significant, the assured minimum strength will likely be too low for practical purposes; however, the number of surviving specimens whose strength falls below the peak proof stress will be typically small and, furthermore, calculable. If, on the other hand, fatigue is too severe, the surviving population may become indistinguishable from the initial one, and the only result of the proof test will be to break some specimens. Such a test has no rational justification.

Practical Applications

Proof testing has been used frequently for improving the statistical strengths of components, and the results of numerous applications have been published in the literature. Proof testing of space shuttle windows[7] and porcelain insulators[8] has been reported by Wiederhorn and his colleagues. Optical-glass fibers were tested by Tariyal and Kalish.[9] Numerous ceramics for heat-engine applications were also proof tested, among these hot-pressed silicon nitride by Wiederhorn and Tighe[10] and sintered α-SiC by Srinivasagopalan et al.[11] In a more recent study by Kamiya and Kamigaito,[12] proof testing was suggested for use against thermal fatigue.

The best illustration of the effects of proof testing can be found among the studies of soda-lime glass. Figures 3 and 4 are Weibull plots obtained by Ritter et al.[13,14] for the strength of proof-tested soda-lime glass. Figure 3 shows the initial and after-proof strength distributions from tests performed in liquid nitrogen. These tests were essentially fatigueless, and it is evident that the prediction follows the after-proof data quite well. Types I, II, and III designate no fatigue, fatigue on loading only, and fatigue on both loading and unloading, respectively. Figure 4 shows the results of proof testing soda-lime glass in laboratory air. This figure also includes the same inert strength distribution shown in Fig. 3. Again, the prediction follows the data within the experimental fluctuations. These figures underscore the essential soundness of the proof-test theory for at least those applications in which fatigue is not too severe.

Beyond the possibility of excessive fatigue during proof testing, the proof-test theory is inapplicable under some other circumstances. This does not necessarily mean that proof testing in these cases will result in no benefits, but simply that the benefits are unpredictable. For example, proof testing has been shown to be unpredictable when the crack-growth law is more complicated than that in Eq. (1) (soda-lime glass in heptane, for instance).[14] Under such conditions, the after-proof strength distribution can be worse than the initial distribution, and proof testing clearly has no value at all.

Proof testing is also ineffective when the strength-controlling flaw population changes in service. In this case, proof testing may truncate the after-proof distribu-

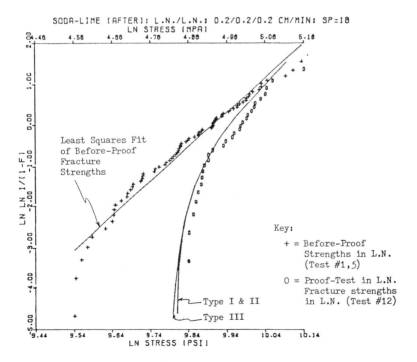

Fig. 3. Weibull plot of initial and after-proof strength distributions for soda-lime glass proof tested in liquid nitrogen (Ref. 13).

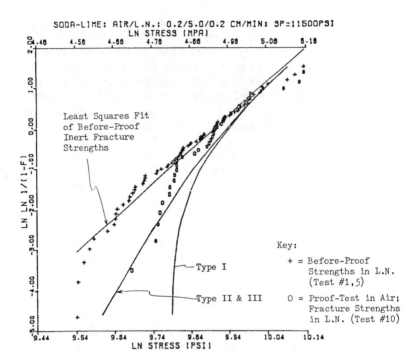

Fig. 4. Weibull plot of initial and after-proof strength distributions for soda-lime glass proof tested in humid air (Ref. 13).

tion, as expected, but a minimum strength is not ensured, since failure in service is not controlled by the proof-tested flaws. Such a situation is shown in Fig. 5: Here, the initial distribution and two after-proof distributions are shown for hot-pressed silicon nitride.[15] Specimens for one of the after-proof distributions were heated to 1000°C for 10 minutes; for the other distribution they were proof tested prior to heat treatment. The two after-proof distributions are indistinguishable from each other, and the proof-tested specimens certainly do not follow the predicted curve (dashed line). In this case, oxidation during heat treatment increased the specimen strength, thus obliterating the original flaw population and, hence, the beneficial effects of proof testing as well. This example vividly demonstrates that if the flaw population changes in service proof testing may have no beneficial effect whatsoever.

Conclusions

Proof testing is a widely used, viable tool for increasing product reliability of many glass and ceramic components. It is, however, a technique whose practical benefits can be diminished by fatigue, unfortunately a rather common strength degradation phenomenon in glass and ceramic materials. To realize the greatest benefits from proof testing, good "proof-test control" must be maintained: Fatigue during the test must be minimized, either by providing a dry, inert environment or by using fast loading and very fast unloading rates. If fatigue is eliminated, the after-proof distribution becomes truncated; hence, a minimum strength is ensured, and furthermore the strength distribution also substantially improves. In fatigueless cases, the proof-test theory predicts the experimental results very well. When

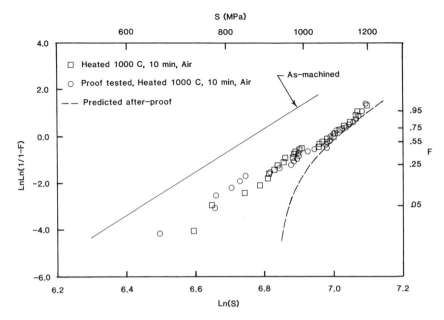

Fig. 5. Weibull plot of heat-treated and proof-tested/heat-treated hot-pressed silicon nitride (Ref. 15).

fatigue becomes significant, one can no longer ensure a minimum strength for the survivors, although the strength of the population as a whole still may be substantially improved by proof testing. When very severe fatigue conditions exist, proof testing may not be beneficial, and proof-test theory is ill-equipped to predict the outcome. Although eliminating fatigue during proof testing may not be a trivial task in many industrial environments, it may nevertheless be necessary in critical applications, since it is perhaps the only way of ensuring ceramic-component reliability.

References

[1]A. G. Evans and S. M. Wiederhorn, "Proof Testing of Ceramic Materials—An Analytical Basis for Failure Prediction," *Int. J. Fract.,* **10** [3] 379–92 (1974).
[2]A. G. Evans and E. R. Fuller, Jr., "Proof Testing—The Effect of Slow Crack Growth," *Mater. Sci. Eng.,* **19**, 69–77 (1975).
[3]S. M. Wiederhorn, "Reliability, Life Prediction, and Proof Testing of Ceramics"; pp. 635–55 in Ceramics for High Performance Applications. Edited by J. J. Burke, A. E. Gorum, and R. N. Datz. Brook Hill Publishing Co., Chestnut Hill, MA, 1974.
[4]R. J. Charles, "Dynamic Fatigue of Glass," *J. Appl. Phys.,* **29** [12] 1657–62 (1958).
[5]S. M. Wiederhorn, "Subcritical Crack Growth in Ceramics"; pp. 613–46 in Fracture Mechanics of Ceramics, Vol. 2. Edited by R. C. Bradt, D. P. H. Hasselman, and F. F. Lang. Plenum, New York, 1974.
[6]E. R. Fuller, Jr., S. M. Wiederhorn, J. E. Ritter, Jr., and P. B. Oates, "Proof Testing of Ceramics: II. Theory," *J. Mater. Sci.,* **15**, 2275–81 (1980).
[7]S. M. Wiederhorn, A. G. Evans, E. R. Fuller, Jr., and H. Johnson, "Application of Fracture Mechanics to Space-Shuttle Windows," *J. Am. Ceram. Soc.,* **57** [7] 319–23 (1974).
[8]A. G. Evans, S. M. Wiederhorn, "Proof Testing of Porcelain Insulators and Application of Acoustic Emission," *Am. Ceram. Soc. Bull.,* **54** [6] 576–81 (1975).
[9]B. K. Tariyal and D. Kalish, "Mechanical Behavior of Optical Fibers"; pp. 161–75 in Fracture Mechanics of Ceramics, Vol. 4. Edited by R. C. Bradt, D. P. H. Hasselman, and F. F. Lang. Plenum, New York, 1978.
[10]S. M. Wiederhorn and N. J. Tighe, "Proof Testing of Hot-Pressed Silicon Nitride," *J. Mater. Sci.,* **13**, 1781–93 (1978).

[11]S. Srinivasagopalan, M. Srinivasan, and G. W. Weber, "Proof Test Studies in Sintered Alpha Silicon Carbide"; for abstract see *Am. Ceram. Soc. Bull.*, **58** [12] 1203 (1979).

[12]N. Kamiya and O. Kamigaito, "The Possibility of Proof Testing Ceramics Against Thermal Fatigue by Mechanical Stress," *J. Mater. Sci.*, **16**, 828–30 (1981).

[13]J. E. Ritter, Jr., P. B. Oates, E. R. Fuller, Jr., and S. M. Wiederhorn, "Proof Testing of Ceramics: I. Experiment," *J. Mater. Sci.*, **15**, 2275–81 (1980).

[14]J. E. Ritter, Jr., K. Jakus, G. M. Young, and T. H. Service, "Effect of Proof Testing Soda-Lime Glass in Heptane," *J. Am. Ceram. Soc.*, **65** [8] 134–35 (1982).

[15]K. Jakus, J. E. Ritter, Jr., and W. P. Rogers, "Strength of Hot-Pressed Silicon Nitride after High-Temperature Exposure," *J. Am. Ceram. Soc.*, **67** [7] 471–75 (1984).

Control Charts and Their Use in Glass-Production Decisions

T. D. TAYLOR

Clemson University
Dept. of Ceramic Engineering
Clemson, SC   29631

Variations can occur from a variety of sources. Variations are to be expected, yet abnormal variations cause considerable problems with production. Control assist in distinguishing the usual variation from an abnormal variation. Discussion includes constructing and using control charts to monitor glass house variables.

Perhaps one of the more critical challenges for a plant (or laboratory) manager is instructing employees when to react to an apparent change in glass house conditions and when to leave things alone. How does one distinguish between normal operation and an upset in the process? Control charts are at least a partial solution to the problem. This paper shows how to construct and use control charts.

What Are Control Charts?

A control chart is a record of the history of a process parameter; superimposed on it are control limits indicating whether this parameter is "in control" (within normal limits) or "out of control" (undergoing an abnormal fluctuation). These limits are determined by the actual variations found in the process parameter itself. When the process fluctuates beyond these limits, a change in the process has occurred and action is required.

The value of a process parameter is usually determined from the average of several of its measurements, a group called a sample. Calculating the amount of variation in these measurements is also useful. Perhaps the simplest measure of this variation is the range: the difference between the largest and the smallest value.

It should be pointed out that there are many types of control charts, useful for various purposes.[1] The control charts discussed here, however, are very simple and suitable for use by production personnel: Only the plotting of data and the calculation of averages and ranges are required.

Table I is a simulated series of glass density measurements. Assume that these samples were taken on 50 consecutive days from one tank. The first five columns are the individual measurements for each sample; the sixth column is the average value for each sample, and the seventh the range for each sample.

Figure 1 plots the average density on each day. This chart is quite worthless, since there is no means for judging the significance of the fluctuations.

Figure 2 is a similar chart showing the range of each sample. Again, what is the significance of the fluctuations? What is a normal range? When is action warranted?

The objective of a control chart is to establish criteria for determining abnormalities. It has been found that variations of three standard deviations from the average are a practical limit for detecting process upsets. Variations of this magnitude occur only three times in a thousand measurements for a process that is

Table I. Hypothetical Density Data

Sample No.	1	2	3	4	5	Avg.	Range
1	2.5004	2.5000	2.4997	2.4993	2.4997	2.4998	0.0011
2	2.4996	2.4997	2.4999	2.4996	2.4994	2.4996	0.0005
3	2.5006	2.5003	2.4999	2.5001	2.5001	2.5002	0.0007
4	2.4995	2.4997	2.4999	2.4998	2.4994	2.4997	0.0005
5	2.5003	2.5008	2.5007	2.5003	2.5012	2.5007	0.0009
6	2.5004	2.5006	2.4998	2.5005	2.5001	2.5003	0.0008
7	2.4993	2.4990	2.4994	2.4989	2.4991	2.4991	0.0005
8	2.4995	2.4991	2.4993	2.4995	2.4997	2.4994	0.0006
9	2.4993	2.4999	2.4999	2.4991	2.4997	2.4996	0.0008
10	2.4999	2.5000	2.5003	2.4998	2.4995	2.4999	0.0008
11	2.4990	2.4997	2.4991	2.4991	2.4998	2.4993	0.0008
12	2.4997	2.4995	2.4997	2.5000	2.5001	2.4998	0.0006
13	2.5005	2.5002	2.5007	2.5004	2.5000	2.5004	0.0007
14	2.4999	2.4997	2.4993	2.4998	2.5005	2.4998	0.0012
15	2.4999	2.4996	2.4993	2.5003	2.4999	2.4998	0.0010
16	2.5011	2.5013	2.5009	2.5005	2.5013	2.5010	0.0008
17	2.5008	2.5000	2.5003	2.5004	2.5005	2.5004	0.0008
18	2.4994	2.5000	2.4997	2.4996	2.4996	2.4997	0.0006
19	2.5006	2.5001	2.5002	2.5002	2.5000	2.5002	0.0006
20	2.5000	2.5001	2.4997	2.4998	2.5000	2.4999	0.0004
21	2.5000	2.4994	2.5004	2.4997	2.5003	2.5000	0.0010
22	2.5001	2.4996	2.4990	2.5000	2.4994	2.4996	0.0011
23	2.5003	2.5000	2.4997	2.5000	2.5000	2.5000	0.0006
24	2.4993	2.4995	2.4993	2.4993	2.4991	2.4993	0.0004
25	2.4995	2.4996	2.4998	2.4993	2.4999	2.4996	0.0006
26	2.5003	2.4998	2.5000	2.5005	2.5010	2.5003	0.0012
27	2.4999	2.4999	2.5000	2.5002	2.4999	2.5000	0.0003
28	2.4997	2.4992	2.4997	2.5000	2.4996	2.4996	0.0008
29	2.4995	2.4998	2.4999	2.4998	2.4994	2.4997	0.0005
30	2.4997	2.4998	2.5003	2.5003	2.5002	2.5001	0.0006
31	2.5001	2.4992	2.4994	2.5002	2.4998	2.4997	0.0010
32	2.5000	2.4998	2.5000	2.4996	2.5000	2.4999	0.0004
33	2.5008	2.4999	2.5004	2.5008	2.5006	2.5005	0.0009
34	2.5004	2.4997	2.5002	2.4998	2.5003	2.5001	0.0007
35	2.4990	2.4995	2.4995	2.4996	2.4991	2.4993	0.0006
36	2.4990	2.4995	2.4989	2.4991	2.4995	2.4992	0.0006
37	2.4991	2.4992	2.4999	2.4992	2.4994	2.4994	0.0008
38	2.5002	2.5002	2.5002	2.4995	2.5003	2.5001	0.0008
39	2.5004	2.5003	2.5001	2.5001	2.4996	2.5001	0.0008
40	2.5002	2.5001	2.4994	2.4999	2.4999	2.4999	0.0008
41	2.5005	2.5005	2.5002	2.5009	2.5003	2.5005	0.0007
42	2.5001	2.4999	2.5003	2.4993	2.4999	2.4999	0.0010
43	2.5001	2.4993	2.4996	2.4996	2.4995	2.4996	0.0008
44	2.5001	2.5002	2.5001	2.5000	2.5005	2.5002	0.0005
45	2.5004	2.4999	2.5003	2.5003	2.5004	2.5003	0.0005
46	2.4996	2.4999	2.4997	2.4992	2.4999	2.4997	0.0007
47	2.5006	2.5001	2.5003	2.5006	2.5006	2.5004	0.0005
48	2.5001	2.5005	2.5003	2.5001	0.5002	2.5002	0.0004
49	2.5007	2.4998	2.5003	2.4998	2.5005	2.5002	0.0009
50	2.5002	2.5003	2.5003	2.5008	2.4998	2.5003	0.0010
				Mean Values:		2.4999	0.000724

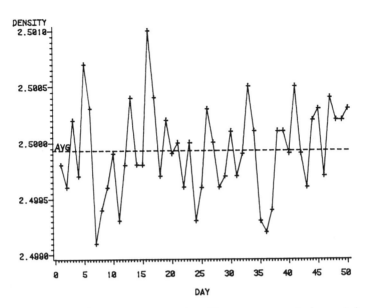

Fig. 1. Glass density averages vs day (5 measurements/sample).

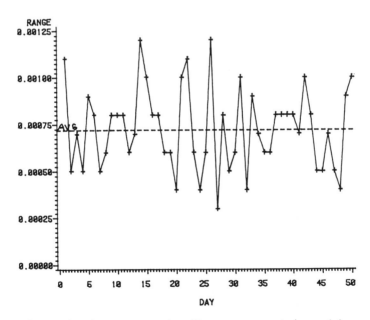

Fig. 2. Glass density range vs day (5 measurements/sample).

Table II. Factors for Computing Control-Chart Limits

Sample Size	Chart for Averages	Chart for Ranges	
		Lower	Upper
2	1.880		3.267
3	1.023		2.575
4	0.729		2.282
5	0.577		2.115
6	0.483		2.004
7	0.419	0.076	1.924
8	0.373	0.136	1.864
9	0.337	0.184	1.816
10	0.308	0.223	1.777
11	0.285	0.256	1.744
12	0.266	0.284	1.716
13	0.249	0.308	1.692
14	0.235	0.329	1.671
15	0.223	0.348	1.652
16	0.212	0.364	1.636
17	0.203	0.379	1.621
18	0.194	0.392	1.608
19	0.187	0.404	1.596
20	0.180	0.414	1.586
21	0.173	0.425	1.575
22	0.167	0.434	1.566
23	0.162	0.443	1.557
24	0.157	0.452	1.548
25	0.153	0.459	1.541

actually under control. In other words, when the process does fluctuate beyond these limits, the changes may almost always be attributed to a real shift in the process. In the case of density, the variations could indicate a weighing problem, a change in tank currents, or even a problem with the density measurement itself.

Constructing Average and Range Control Charts

Control charts may be effectively constructed only after the limits of the process are established. This is done by simply measuring at least 30 samples. In the present example 50 samples were used as a basis for establishing the charts, each sample consisting of 5 measurements. The number of measurements in any sample depends on the type of measurements being made and the preferences of the personnel involved.

These preliminary measurements establish the control limits. The procedure is best explained by a demonstration:

1. The overall average is found first. One can simply take the sum of every measurement and divide by the total number of measurements, or take the average of the sample averages. The calculated value becomes the central line on the control chart for the averages.

2. The average of all the ranges is found by taking the average of the individual sample ranges. The accuracy of this calculation is very important, since the control limits are based on this value.

3. Table II lists factors used with the range average for establishing the control limits. To establish the upper and lower limits for the density data, locate the 0.577 across from the 5 in the sample-size column. The upper limit on sample averages is established by adding the product of the chart factor for averages (0.577) and the range average (0.000724) to the overall average. Similarly, the lower limit is found by subtracting this product from the overall average.

4. Table II also shows factors necessary for constructing range control charts. These factors are based on sample size (in our case, 5). Unlike the control chart for averages, however, this chart shows a factor for the upper and also the lower limit. The lower limit for a given sample size is the product of the lower range factor (0) and the range average (0.000724). The upper limit is found in a similar manner: in our case, the product of 0.000724 and 2.115.

The above calculations have produced upper and lower limits for both the average and the range control charts. All that remains is to draw these lines on charts and begin plotting sample averages and sample ranges; the results are shown in Figs. 3 and 4. Figure 3 indicates that the density average was out of control on at least 14 days. Some effort to correct the situation should have been made on each of those days. The range remained in control during the entire period and no action should have been initiated.

Discussion

Our emphasis has been on using enough samples to obtain an accurate estimate of the data parameters. This is extremely important: Poor estimates create situations in which poor decisions are made. No fewer than 25 samples should be used when constructing a control chart.

For successfully using a control chart, the specimens must be gathered and

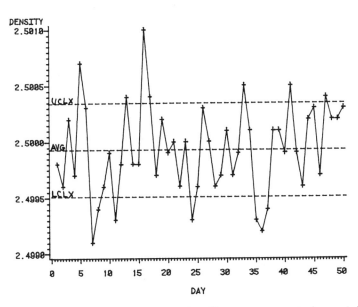

Fig. 3.   Glass density averages vs day (5 measurements/sample), showing upper and lower control limits.

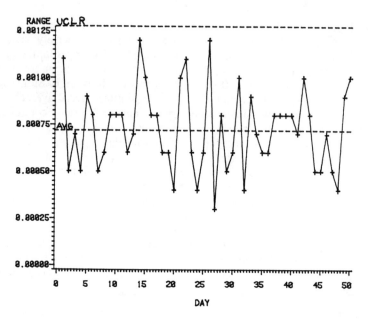

Fig. 4.   Glass density range vs day (5 measurements/sample), showing upper and lower control limits.

measured promptly, then immediately charted. Old samples and data merely indicate what action should have been taken.

The control chart is not a goal-setting tool: Limits are established by the process itself, not by the operator or his supervisor. If these limits are intolerably large, the process must be changed to reduce them. Arbitrarily reducing the limits of variability merely results in more false alarms.

While it is possible to construct control charts with varying sample sizes, the extra work involved in establishing the control limits precludes the simplicity desired of a production tool. In most cases, establishing a constant, practical sample size is better.

Loss of control in a process can be anticipated by constructing secondary control limits only two standard deviations from the line representing the overall average. When a sample average exceeds these narrower limits, the process is watched more intensely. Another approach is to note the number of consecutive samples located on the side of the line representing the overall average. The sample average rarely deviates in the same direction for four consecutive periods: Four consecutive deviations of the same sign strongly indicate a process shift.

Conclusions

Control charts can indicate action necessary to keep a process operating properly. On the other hand, the control chart helps operating personnel avoid overcontrolling a process. Sample averages and ranges are easy to calculate, and control charts involving these simple calculations are more likely to be adopted by plant personnel. It should be emphasized that prompt sample measurement, charting, and action are necessary if this production tool is to be used most efficiently.

Reference

[1]N. L. Johnson and F. C. Leone, Statistics and Experimental Design in Engineering and the Physical Sciences, 2d ed., Vol. I; pp. 340–90. Wiley & Sons, New York, 1977.

Application of Microcomputers to Glass Problems

RICHARD L. LEHMAN

Rutgers University
Department of Ceramics
Piscataway, NJ   08854

This paper traces trends in applying digital computers to numerical glass problems from the development of mainframe computers in the 1960s through the advent of microcomputers. Microcomputers have become sufficiently advanced to be used in solving a wide variety of glass-development and production problems. Major advantages are for small companies where no prior computing capability existed, or for large companies where computing needs are decentralized. Microcomputer software programs developed for batch formulation, raw-material selection, and glass-property estimation are discussed and examples computed for each category.

The potential for applying digital computers in commercial glass development and manufacturing activities is considerable. In fact, nearly all glass manufacturers use either minicomputers or small microprocessor units to control a wide variety of manufacturing processes such as batch weighout, mixing and charging, furnace-combustion control, forming-operation control, and numerous post-forming operations. Most major glass companies also provide access to a large mainframe or minicomputer to assist in computations which solve offline glass-production problems. Examples of traditional applications are batch calculations by simultaneous equations or linear programming, property prediction from standard factors or through structural modeling, and collection and computation of quality-control data. Emerging uses for digital computers include CAD/CAM and similar techniques which use advanced software and peripherals.

In recent years, the introduction of the microcomputer has sharply altered ways in which computing power can be applied to glass manufacturing and development problems. This article reviews the potential impact of microcomputers on commercial glass manufacture and discusses several specific applications in detail.

Microcomputer Development

For practical purposes, applying digital computers to scientific problems began on a large scale in the early 1960s, with development of the IBM 360 series mainframe computers. Although these computers could solve large problems, only the largest companies acquired them due to high cost and the necessary commitment to a major computing facility. Some time-sharing facilities were available, but computer application to glass problems consisted primarily of research-oriented problems at major scientific laboratories, universities, or large corporations. During this same period, of course, great benefit was derived from digital-process control, which displaced many of the traditional analog-control systems for temperature, combustion, and material transport in the modern glass plant.

213

The advent of minicomputers such as the well-known DEC* and VAX[†] systems further promoted the use of computers, since these systems were smaller and less expensive, yet had many of the same capabilities as the mainframe computers of the 1960s. Many intermediate-sized companies found these computers suitable for process control and, in addition, inexpensive enough to justify a dedicated in-house unit for solving numerical problems in research, development, and manufacturing. As such, many glass companies began to expand the application of computers to technical problems that previously had been solved by simpler means or not addressed at all. During this period, the early 1970s, some glass companies developed programs for computerized batch calculations and optimization. Networking and time-sharing arrangements proliferated at the minicomputer level, much as they had with the earlier mainframe computers, and most major glass companies found that these systems met virtually all of their computing needs.

When microcomputers were introduced in the 1970s, their enormous potential impact was not immediately evident, due in part to the limited capability of the early 8-bit processors and the virtually nonexistent technical software. Rapid technological advances and the entrance of large computer companies (IBM, DEC) into the microcomputer market increased the capabilities of the micros, enhanced the credibility of the small machines as bona fide computers, and created the technical software market. The current 16- and 32-bit processors available in microcomputers provide enough speed to be used in rather sophisticated glass problems which were beyond the scope of the early 8-bit machines. Table I summarizes the current status of microcomputer hardware.

Which computers will dominate in the future? This certainly is a difficult question to face, because of the rapid pace of technological development. Two trends, however, presently appear to be developing: enhanced communications capability between computers, especially microcomputers, which permits efficient development of local area networks (LANs), and the continued evolution of microcomputers. In fact, as microcomputers continue to evolve and the functional distinction between micros and minis blurs; minis may eventually disappear and large micros will provide the computing power and LANs the networking associated with today's minicomputers.

Applications

The development of microcomputers into computers capable of solving real glass problems primarily benefits the small glass manufacturer who previously was financially unable to apply digital computers to his problems. Large manufacturers also benefit, however, especially when responsibilities are decentralized and no central computer system or database is necessary. Incentives and applications for microcomputers are summarized in Table II. Of the applications in this table, the two most universally important and interesting are computer-assisted batch calculations and physical-property estimation. Methods for accomplishing each of these tasks will be discussed.

Batch Calculations

The computation technique most suitable for accurately calculating commercial glass batches is linear programming (LP), a technique which previously required large mainframe computers.[1] The LP glass-batch formulation procedure

---

*Digital Equipment Co., Maynard, MA.
[†]Digital Equipment Co.

Table I. Current Capabilities of Microcomputers

Item	Status	Capabilities
Processor	16- and 32-bit	Large matrix operations
Coprocessors	Available	Increased speed
RAM	>500 kB	Large problem size
Storage	Disks, hard and soft	File size
Software	Developing	Numerous technical programs currently available
Peripherals	Assorted	Ease of operation, reporting, communications

Table II. Applications for Microcomputers in Commercial Glass Manufacture

Incentives

Small, self-contained computing facility
Easily dedicated to a specific task
Easy to use, no system JCL or communications commands
Always available (minimal "busy" or "down" time)
Special graphics capabilities (lightpens, mouse, etc.)
Inexpensive

Applications

Batch calculations
Raw-material selection and optimization
Glass-property estimation
CAD/CAM (product design, mold design)
Quality-control data collection and computation
Statistical analysis of data
Heat flow and other general calculations

has numerous advantages: improved accuracy of formulation, especially when many complex raw materials are employed, and the abilities to easily establish tolerance limits for each oxide level and to incorporate physical and chemical parameters such as segregation potential, dust levels, and COD values into the batch-formulation process. Thus it is possible to control many key glass-batch parameters in a single, integrated computational package. Software has been developed which permits batch formulation on microcomputers according to the model presented in Ref. 1. This model enables the definition of many physical and chemical batch specifications, such as glass-oxide percentages of both major components and minor additives, desired segregation and dust level of the mixed batch, and the COD balance. Furthermore, numerous additional quantifiable batch parameters can be incorporated as needed. Batch raw materials are selected and optimized in the standard way, according to a defined objective function.

The computational procedure of the model consists of an LP simplex algorithm which iteratively determines a solution to equality and inequality batch relationships. Application of equality or inequality relationships depends on the individual variables selected: For example, it may be desirable to have the Na_2O

215

Table III. Raw-Material Database Used in Formulating Borosilicate and Soda-Lime-Silicate Glasses*

							Raw Materials[†]							
No. Constraint	1 Sand	2 China	3 Nephel	4 Feld	5 Aplite	6 Lime	7 Slag	8 Anhbor	9 Boracd	10 5MBor	11 BDLime	12 Salt C	13 Sod (M)	14 Sod (S)
1 SiO_2	99.6	44.5	60.7	67.1	63.6	0.4	37.7	0.1	0.0	0.0	0.9	0.0	0.0	0.0
2 Al_2O_3	0.1	39.6	23.3	18.3	22.0	0.1	9.0	0.1	0.0	0.0	0.6	0.0	0.0	0.0
3 CaO	0.0	0.0	0.7	0.4	5.5	54.8	41.0	0.0	0.0	0.0	57.1	0.0	0.0	0.0
4 MgO	0.0	0.0	0.1	0.0	0.0	0.4	9.0	0.1	0.0	0.0	39.2	0.0	0.0	0.0
5 B_2O_3	0.0	0.0	0.0	0.0	0.0	0.0	0.0	69.2	56.5	47.8	0.0	0.0	0.0	0.0
6 Na_2O	0.0	0.1	9.8	3.8	6.0	0.0	0.0	30.8	0.0	21.3	0.0	43.6	58.4	58.4
7 K_2O	0.0	0.1	4.6	10.1	2.6	0.0	0.0	0.0	0.0	0.0	0.0	0.0	0.0	0.0
8 Fe_2O_3	0.0	0.4	0.1	0.1	0.1	0.0	0.3	0.0	0.0	0.0	0.2	0.2	0.0	0.0
9 SO_3	0.0	0.0	0.0	0.0	0.0	0.0	2.6	0.0	0.1	0.1	0.3	56.0	0.0	0.0
10 Seg	0.0	288.0	0.0	130.0	20.0	60.0	105.0	145.0	50.0	170.0	160.0	5.0	5.0	55.0
11 Dust	7.4	100.0	16.0	44.0	3.0	12.0	20.0	15.0	30.0	13.0	33.0	10.0	10.0	13.0
12 Cost	25.00	100.00	70.00	71.00	45.00	39.60	33.40	644.00	632.00	281.00	43.50	215.20	139.60	139.60

*Data for wt% of constraint oxide, except segregation and dust parameters.
[†]Key to raw material abbreviations: China = air-floated china clay; Nephel = nephelene syenite; Feld = feldspar; Lime = limestone; Anhbor = anhydrous borax; Boracd = boric acid; 5MBor = 5-mole borax; BDLime = burned dolomitic limestone; Salt C = salt cake, sodium sulfate; Sod (M) = natural soda ash, mono process; and Sod (S) = natural soda ash, sesqui process.

level at exactly 13.7%, which would dictate an equality statement, whereas any segregation level may be permitted, provided it is less than a given value (inequality). Each parameter is considered individually when setting up the model. Once a solution to the inequality and equality constraints has been established, the solution is modified to optimize the objective function, as permitted by the extent of slackness in the model. When the optimization procedure is complete, a comprehensive report is printed. The software developed by the author runs on a standard IBM PC with 128K RAM, two 320K disk drives, and an 80-column printer.

Two examples illustrate the capability of the formulation program: (1) an alkali borosilicate batch typical of those used in fiberglass wool manufacture and (2) a standard soda-lime glass.

The batch formulation objective for the alkali-borosilicate glass was twofold. Initially, it was desired to calculate a batch with the lowest possible cost consistent with meeting the defined oxide composition and with targeted values for segregation and dust levels. Fourteen raw materials were identified as suitable for the given composition; the property/raw-material matrix, as printed by the program, is given in Table III. The specified glass properties and the properties calculated by the program are given in Table IV. Property specifications are met almost exactly for equality constraints; minor differences are due to the single-precision limitations of the program. The precision limitations can cause problems whenever property values range over several orders of magnitude, as with COD. Anticipated slackness is evident in the three inequality properties. The batch composition, which used 8 of the 14 raw materials, is given in Table V. The object function, representing the optimization parameter, was total batch cost and achieved a minimum value of $72.35 per ton of glass.

In a variation on this first example, the same batch was calculated, but with the objective of minimizing particle-segregation tendency rather than cost. Output data for the minimum-segregation batch are given in Tables VI and VII. The segregation potentials of the two batches differ substantially: The cost-minimized batch has a segregation index of 72.3 and the segregation-minimized batch an index of 38.1. The lower segregation index results from selecting raw materials,

Table IV. Specified and Calculated Properties of an Alkali-Borosilicate Glass*

| No. | Property | Properties of the Glass | | | Equality Type |
		Specified	Calculated	Difference	
1	SiO_2	58.50	58.48	0.02	=
2	Al_2O_3	4.25	4.24	0.01	=
3	CaO	15.50	15.50	0.00	=
4	MgO	5.50	5.50	0.00	=
5	B_2O_3	3.50	3.50	0.00	=
6	Na_2O	11.00	11.00	0.00	=
7	K_2O	1.00	1.00	0.00	=
8	Fe_2O_3	0.15	0.09	0.06	<
9	SO_3	0.60	0.60	-0.00	=
10	Seg	100.00	72.32	27.68	<
11	Dust	20.00	17.92	2.08	<

*Formulation subject to cost minimization.

217

Table V. Batch Formula for Cost-Minimized Alkali-Borosilicate Glass

	Batch Composition	
No.	Batch Component	Weight
1	Sand	42.63
4	Feld	9.37
5	Aplite	1.91
6	Lime	2.12
7	Slag	22.25
10	5Mbor	7.32
11	Bdlime	8.88
14	Sod (S)	15.35
Total batch weight		109.86
Loss on ignition		9.95
Net glass weight		99.91
Object function optimum		72.35 (minimum cost, $/ton glass)

Table VI. Specified and Calculated Properties of an Alkali-Borosilicate Glass*

		Properties of the Glass			
No.	Property	Specified	Calculated	Difference	Equality type
1	SiO_2	58.50	58.50	−0.00	=
2	Al_2O_3	4.25	4.25	−0.00	=
3	CaO	15.50	15.54	−0.04	=
4	MgO	5.50	5.51	−0.01	=
5	B_2O_3	3.50	3.50	0.00	=
6	Na_2O	11.00	11.00	−0.00	=
7	K_2O	1.00	1.00	−0.00	=
8	Fe_2O_3	0.15	0.06	0.09	<
9	SO_3	0.60	0.61	−0.01	=
10	Dust	20.00	16.95	3.05	<

*Formulation subject to segregation minimization.

such as nephelene syenite, limestone, boric acid, and monoprocess soda ash, that are close in particle-size to the sand. Of course, using some low-segregation materials, such as boric acid rather than 5-mole borax, means that a large economic price must be paid: The minimum batch cost was $72.35 per ton of glass, compared with $98.80 per ton for the minimum-segregation batch. The severity of the segregation problem must then be weighed against the economic penalty for correcting it when making a final decision. Aplite and slag, noted as poor particle-size fits with the sand, were thus eliminated.

The second example of glass-batch formulation using linear programming considers a standard soda-lime composition such as may be used for flat glass or containers. The same raw materials were used as in the first example, except that the boron-containing materials were dropped. The desired glass properties, batch formulation, and calculated values are given in Table VIII. The objective function

Table VII. Batch Formula for Segregation-Minimized
Alkali-Borosilicate Glass

	Batch Composition	
No.	Batch Component	Weight
1	Sand	47.24
3	Nephel	15.28
4	Feld	2.93
6	Lime	13.68
9	Boracd	6.19
11	BDLime	13.86
12	Salt cake	0.99
13	Sod (M)	15.33
Total batch weight		115.55
Loss on ignition		15.59
Net glass weight		99.96
Object function optimum		38.07
Batch cost, $/ton glass		98.80

Table VIII. Properties and Batch Formula for Cost-Minimized
Soda-Lime Glass

		Properties of the Glass			
No.	Property	Specified	Calculated	Difference	Equality Type
1	SiO_2	68.00	68.00	−0.00	=
2	Al_2O_3	2.35	2.35	−0.00	=
3	CaO	11.50	11.53	−0.03	=
4	MgO	2.10	2.10	0.00	=
5	Na_2O	14.00	14.00	−0.00	=
6	K_2O	1.00	1.00	−0.00	=
7	Fe_2O_3	0.15	0.05	0.10	<
8	SO_3	0.90	0.90	−0.00	=
9	Seg	100.00	33.55	66.45	<
10	Dust	20.00	15.03	4.97	<

	Batch Composition	
No.	Batch Component	Weight
1	Sand	59.66
4	Feld	9.90
6	Lime	13.09
7	Slag	4.81
8	BDLime	4.11
9	Sod (M)	22.30
11	Salt cake	1.36
Total batch weight		115.26
Loss on ignition		15.31
Net glass weight		99.94
Object function optimum		64.61 (batch cost, $/ton glass)

was used to minimize cost and returned a value of $64.61 per ton of glass. In a second run, the objective function was used to minimize Fe_2O_3 levels, as shown in Table IX. The major effect of this change was to eliminate the slag, which is relatively high in iron, replacing it with the more expensive raw material nephelene syenite. This replacement resulted in slightly lower Fe_2O_3 levels (0.049–0.047%) and a slightly more expensive batch ($64.61–$65.43). Again, the trade-off of higher cost for better color is a judgment decision which must be made by management.

Property Prediction

It has long been known that many properties of commercial silicate glasses depend in a nearly linear way on composition.[2-5] Viscosity, density, thermal expansion, chemical durability, electrical conductivity, and optical and other properties have been related, through extensive experimentation and correlation analysis, to glass composition. It is also well known that the ability to predict the physical and chemical properties of glass from its chemical composition alone is very useful in research and manufacturing. For glass research aimed at developing

Table IX. Properties and Batch Formula for Fe_2O_3-Minimized Soda-Lime Glass

No.	Property	Properties of the Glass			Equality Type
		Specified	Calculated	Difference	
1	SiO_2	68.00	68.00	−0.00	=
2	Al_2O_3	2.35	2.35	−0.00	=
3	CaO	11.53	11.50	−0.03	=
4	MgO	2.10	2.10	−0.00	=
5	Na_2O	14.00	14.00	0.00	=
6	K_2O	1.00	1.00	−0.00	=
7	Fe_2O_3	0.15	0.04	0.11	<
8	SO_3	0.90	0.91	−0.01	=
9	Seg	100.00	30.00	70.00	<
10	Dust	20.00	14.64	5.36	<
11	Oxide hi	101.00	99.93	1.07	<
12	Oxide lo	99.00	99.93	−0.93	>

No.	Batch Composition	
	Batch Component	Weight
1	Sand	60.61
3	Nephel	2.82
4	Feld	8.61
6	Lime	15.54
8	BDLime	5.19
9	Sod (M)	21.75
11	Salt cake	1.57

Total batch weight	116.12
Loss on ignition	16.19
Net glass weight	99.93
Object function optimum	0. 04
Batch cost, $/ton glass	65.43

new glasses with specific properties and processing characteristics, a knowledge of property/composition relationships is helpful in directing the formulation experiments. In production, similar knowledge can be used for estimating the compositional cause of specific deviations from physical or chemical property specifications.

Research laboratories and major glass companies have long used large computers to model property/composition relationships. Microcomputers now can program and run many property/composition models, but large models including numerous properties and/or based on molecular models of the glass structure are still beyond the range of most microcomputers. For relatively simple relationships, however, reasonably accurate property/composition relationships can be calculated in convenient times on many microcomputers.

From a commercial-glass viewpoint, the two most important glass properties are viscosity and density. A program has been developed for the IBM PC which calculates the viscosity and density of a wide range of commercial glasses based on simple property/composition relationships. Not requiring the large memory necessary for the linear programming in batch calculations, this program was written to use minimal RAM, permitting its subsequent use on even the smallest microcomputer. The viscosity model was based on the Fulcher equation, with coefficients calculated from compositional factors.[5] Calculating the viscosity of NBS-viscosity standard 710 glass and several other reference glasses[6,7] demonstrates good agreement, as illustrated in Fig. 1. Density factors have been reported by many authors. The factors most suitable over a wide range of commercial

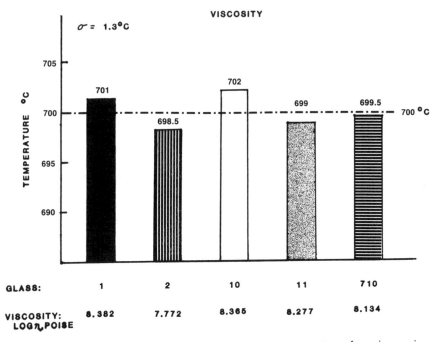

Fig. 1. Comparison of calculated vs known temperature for given viscosity values of several standard glasses. Known temperature is 700°C in all cases.

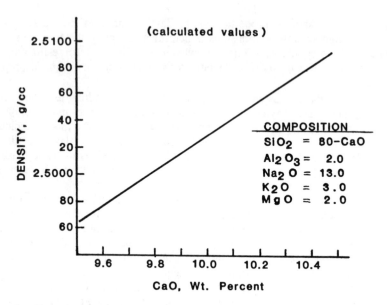

Fig. 2. Density-prediction model applied to calculating glass densities over a range of CaO values for soda-lime-silicate glass.

glasses and with minimum calculation were determined by Sun et al.[8] As an example of the data generated by these density-prediction equations, the calculated effect of small calcium substitutions for silica on the density of a soda-lime-silica glass are given in Fig. 2. For well-annealed glasses, the standard deviation between calculated and experimental values is ≈ 0.001 g/cm^2.

A basic program using both the viscosity and the density calculations discussed previously occupies less than 20 kB of storage and is easily executed in the lower 64 kB of RAM memory accessible to the BASIC compiler.

Summary and Conclusion

The application of digital computers to glass problems has progressed with the development of increasingly more accessible hard- and software. The present availability of 16- and 32-bit microcomputers with large RAM areas and a wide assortment of peripheral devices has enabled desk-top computers to solve many glass problems previously confined to mainframe machines or minicomputers. This evolution of computing ability enables freedom from centralized computing systems for applications not requiring access to a common database, and allows small companies to gain computer capability where it was previously too expensive. Batch formulation and raw-material selection, and glass-property prediction are two areas in which microcomputer application at the manufacturing level can be beneficial. Simple programs that run on the IBM PC have been developed for batch calculations by linear programming and glass-property calculation from prediction equations. The LP program requires 120 kB, but the simpler property-prediction program operates within the lower memory accessed by the BASIC compiler.

References

[1]R. L. Lehman, "A New Batch Formulation Method," *Glass Ind.*, **64** [12] 24–28 (1983).
[2]G. W. Morey, Properties of Glass. Reinhold Publishing Co., 1938.

[3]K. C. Lyon, "Prediction of the Viscosities of 'Soda-Lime' Silica Glasses," *J. Res. Natl. Bur. Stand., A,* **78** [4] 497 (1974).

[4]J. P. Poole and M. Gensamer, "Systematic Study of Effect of Oxide Constituents on Viscosity of Silicate Glasses at Annealing Temperatures," *J. Am. Ceram. Soc.,* **32** [7] 220 (1949).

[5]T. Lakatos, "Viscosity-Temperature Relations in Glasses Composed of SiO_2-Al_2O_3-Na_2O-K_2O-CaO-MgO-BaO-ZnO-PbO-B_2O_3," *Glassteck. Tidskrift,* **31** [3] 51 (1976).

[6]C. L. Babcock and D. A. McGraw, "Application of Glass Properties Data to Forming Operations," *Glass Ind.,* **38** [3] 137–61 (1957).

[7]H. A. Robinson and C. A. Peterson, "Viscosity of Recent Container Glass," *J. Am. Ceram. Soc.,* **12** [5] 129–38 (1944).

[8]K. H. Sun, R. M. Welsch, and M. L. Huggins, "Representation of Density and Optical Properties of Ternary Silicate Glass Systems," *J. Am. Ceram. Soc.,* **29** [3] 59 (1946).

Author Index

Subject Index